VALUATION OF MARINAS

By John A. Simpson, MAI

**APPRAISAL
INSTITUTE®**

875 North Michigan Avenue
Chicago, Illinois 60611
www.appraisalinstitute.org

Reviewers: Fred DePascale, SRPA, SRA
Daniel Geary, SRA
Nicholas Haddad, MAI
Thomas Motta, MAI, SRA
John Schwartz, MAI
George Silver, MAI, SRA

Vice President, Educational
Programs and Publications: Sean Hutchinson
Director, Content Development
and Quality Assurance: Margo Wright
Manager, Book Development: Stephanie Shea-Joyce
Manager, Design/Production: Julie B. Beich
Editor: Linda Willet

For Educational Purposes Only

The material presented in this text has been reviewed by members of the Appraisal Institute, but the opinions and procedures set forth by the author are not necessarily endorsed as the only methodology consistent with proper appraisal practice. While a great deal of care has been taken to provide accurate and current information, neither the Appraisal Institute nor its editors and staff assume responsibility for the accuracy of the data contained herein. Further, the general principles and conclusions presented in this text are subject to local, state, and federal laws and regulations, court cases, and any revisions of the same. This publication is sold for educational purposes with the understanding that the publisher is not engaged in rendering legal, accounting, or other professional service.

Nondiscrimination Policy

The Appraisal Institute advocates equal opportunity and nondiscrimination in the appraisal profession and conducts its activities in accordance with applicable federal, state, and local laws.

Printed in the United States of America

Library of Congress Cataloging-in-Publication Data
Simpson, John A. (John Andrew), 1962.
  Valuation of marinas / by John A. Simpson.
  p. cm.
  Includes bibliographical references (p.).
  ISBN 0-92-215451-1
  1. Marinas—Valuation—United States.   2. Yacht clubs—Valuation—United States.
3. Boatyards—Valuation—United States.   I. Title.
  VK369.5 .S54   1998
  333.33'9—ddc21

                                                                      98-40571
                                                                      CIP

Table of Contents

Chapters

Readers of this text may be interested in the following books available from the Appraisal Institute:

- *The Appraisal of Real Estate,* eleventh edition
- *The Dictionary of Real Estate Appraisal,* third edition
- *A Guide to Appraising Recreational Vehicle Parks* by Robert S. Saia, MAI
- *Environmental Site Assessments and Their Impact on Property Value: The Appraiser's Role* by Robert V. Colangelo, CPG, and Ronald D. Miller, Esq.
- *Golf Courses and Country Clubs: A Guide to Appraisal, Market Analysis, Development, and Financing* by Arthur E. Gimmy, MAI, and Martin E. Benson, MAI

Foreword

Marinas are located on oceans, bays, rivers, and lakes throughout the world. Because marinas are atypical properties, they require a different type of analysis. The management of a marina is highly specialized, site characteristics are unusually important, and a wide range of franchises can create business value. To address these and other factors, an appraiser must have specialized knowledge of the marina industry.

Valuation of Marinas investigates the unique nature of marinas and the problems encountered in their appraisal. It is intended as a learning tool for entry-level appraisers unfamiliar with marina valuations and as supplemental information for experienced appraisers. The book begins with background information including definitions, a brief investment history, and a description of the factors that affect marinas. Readers learn about site differences, building differences, and business elements that distinguish marina properties from other types of properties. These include location factors, riparian rights, flood zone classification, the presence of wetlands and marshes, and dry rack storage. The author describes the four traditional tests of highest and best use and the three traditional approaches to value in relation to marinas.

In the final chapter, the author demonstrates how to put the concepts in the book to practical use with a case study based on an actual assignment. *Valuation of Marinas* is a valuable resource for any appraiser's library. The detailed descriptions of traditional tests and approaches, list of environmental contacts, financial tables, and real-world case study walk appraisers through the valuation of these unique properties.

Joseph R. Stanfield, Jr., MAI, SRA
1998 President
Appraisal Institute

Acknowledgments

No dedication would be complete without thanks to my business partners, Eileen Chmielewski and Colleen Courtney-Morrison, IFA, GAA, for their friendship, understanding, and patience. Their comprehensive reviews helped me to complete the text on time. A special thank you goes to my mother, Edna Simpson, for her support while I was writing this book and throughout my career. In addition, I would like to thank the staff at the Appraisal Institute and the manuscript reviewers for their help and insight.

John A. Simpson, MAI, CCIM, IFAS

About the Author

John Simpson, MAI, CCIM, IFAS, is a partner in Total Real Estate Services, an appraisal, brokerage, and property management company located in Crofton, Maryland. Simpson authored the Appraisal Institute's *Property Inspection* book as well as a myriad of articles in *The Appraisal Journal*. He earned a bachelor of arts in business administration and a bachelor of science in management from Rutgers University as well as a master of business administration in real estate and management from Temple University. Simpson can be reached at john@totalrealestate.com.

CHAPTER ONE

An Introduction to Marinas

The Dictionary of Real Estate Appraisal defines a marina as "a boat basin that provides dockage and other services to pleasure craft."[1] Although this is an adequate definition, it does not describe the wide range of activities and services provided by modern marina facilities.

The characteristics of a marina can affect the types of services it offers. Marinas with deep water slips can provide yacht mooring or access to commercial fishing boats. Those with shallow water access may be restricted to pleasure crafts of various sizes.

Marinas can offer a wide range of utilities to boaters including water, sewage removal, gasoline, electricity, and propane gas. Services to boaters may include restaurants, bait and tackle shops, restrooms, boat repair, boat cleaning, and others. Many of the services provided are synergistic—e.g., a boat sale operation may offer slips to potential purchasers until they can find a more permanent rental, and/or the operation may offer repair services with established brand-name parts.

Some marinas are a source for recreation and events, like the many yacht clubs found along desirable waterways. Others are part of a large resort or condominium project which offers slips in addition to real estate. The activities and services provided allow a marina to create a unique niche and can significantly impact the value of a property. In spite of all the services offered, most marinas can be classified into three basic categories.

TYPES OF MARINAS

The three basic categories of marinas are recreational marinas, yacht clubs, and boatyards. Even though some of the services or activities offered by these marinas overlap, each type of marina has characteristics that differentiate it from the other types.

A recreational marina is a facility that caters to boaters who use their boats for pleasure or only incidental, noncommercial activities. A yacht club is a large recreational marina that usually has one or more large buildings used by members of the facility. A boatyard is a marina that offers significant repairs and services for larger boats, usually yachts and commercial fishing vessels.

The public's image of a marina is most often the recreational marina. In the past, this type of marina consisted of slips, a gasoline pump, and a management office, but the increasing needs of boaters have prompted recreational marinas

---

1. Appraisal Institute, *The Dictionary of Real Estate Appraisal,* 3d ed. (Chicago: Appraisal Institute, 1993), 219.

In addition to the common gasoline pump shown here, marinas can offer utilities such as water, sewage removal, electricity, and propane gas, and services such as restaurants, bait and tackle shops, restrooms, boat repair, boat cleaning, and more.

to offer services such as restaurants, boat supply stores, on-site storage facilities, and dock utilities.

Yacht clubs offer more services than recreational facilities do; however, yacht clubs may be highly restrictive or expensive to join. Yacht club services include at least one large meeting area, a boat storage facility, and slips. Social activities are emphasized, and recreational programs may include tennis clubs and baseball teams. Members pay a yearly fee for using the privileges of the yacht club.

Boatyards differ from the other two types of marinas in that they are primarily commercial facilities. On-site services may include boat dealerships, repair yards, and other related businesses. The depth of slips and the services offered with slips can vary greatly. Operations performed and services offered at boatyards are usually interrelated. For example, boatyards may lease land to businesses that construct buildings and provide services that complement those provided by the boatyard.

WHAT MAKES MARINAS UNIQUE?

Marinas are different from most commercial properties in several ways. The value of a marina's land and site improvements, which includes the bulkheads and piers, often outweighs the value of the building improvements in all but the largest commercial projects. Another important difference is zoning. Many marinas are located in areas zoned for "marine-type" use, usually prohibiting commercial ventures not connected to the marina's operation or services. This restrictive zoning may allow few alternative uses for the site.

Income and quality of materials are additional areas of difference. In the northern areas of the country, boating is a seasonal activity, and this causes pleasure boat marina incomes to fluctuate during the year. Another factor is the location of many marina sites within severe flood zones. Since improvements can easily be flooded, lower-quality construction materials and finishes are sometimes used for building.

More specialized knowledge is needed to operate a marina than many types of general commercial properties. This results in higher management allocations and more expensive insurance. Incomes are more diverse, and utility expenses are more varied. Other expenses can include riparian rights payments and dredging (which requires governmental approval).

OWNERSHIP AND INVESTMENT POTENTIAL OF MARINAS

At the present time, most marinas are managed by small business concerns and family owners. This pattern of management is starting to change for several reasons. One is the acquisition of small marinas by large marina operations, and the other is the recession of the 1990s.

During most of the 1990s, marinas were perceived as a poor investment. The recession of the 1990s resulted in marina bankruptcies throughout the United States. As a result of foreclosures, financial institutions found themselves with a specialized form of real estate that was unfamiliar to them and difficult to operate. These institutions hired dockmasters and knowledgeable companies to manage the properties until the properties were liquidated, usually at below-market prices. Appraisers tended to use higher capitalization rates due to the "softness" in the market. Some appraisers did not fully understand the unique cash flow inherent in marina operations, the costs to dredge, and the expenses of remodeling a dilapidated facility.

For the most part, this situation seems to have reversed itself. Slip vacancies have declined and rates have begun to rise, making marinas more attractive investments. However, the specialized nature of marinas remains. Capitalization rates vary widely, depending on the businesses and services offered at the marina and their inter-relationships. It is difficult to derive the correct capitalization rate by simply looking at the location of the facility and the condition of the improvements. Such factors as the presence and types of utilities, water depth, siltation patterns, proximity to access points leading to the ocean, marina equipment included in the sale, and the relatively small number of potential investors in this market make it difficult for the appraiser to observe capitalization rate differences. To discover the appropriate capitalization or discount rates for the subject, an appraiser needs to speak with participants in the market. Some of the obstacles affecting the growth of marinas appear in Table 1.1 at the end of this chapter.

FACTORS AFFECTING MARINAS

Marinas are subject to the same macroeconomic and microeconomic variables that affect other commercial properties. Macroeconomic variables include the political

The value of a marina's land and site improvements, such as this fixed pier, often outweighs the value of the building improvements in all but the largest commercial projects.

environment and its effect on the availability of financing; socioeconomic forces such as population growth, population aging, the family unit, and disposable income; and other factors that affect the supply of and demand for marinas. Supply and demand are also affected by microeconomic factors specific to marinas. Local market forces, legal and environmental constraints, competitors, and marina characteristics such as quality of management, site and building characteristics, business elements, potential for dockominium approval, and similar factors have a more isolated effect. Even the price of gasoline can affect marinas. During the process of analyzing a marina, an appraiser must be aware of both the macro-economic and microeconomic forces at work.

Political Environment

Some of the macroeconomic and microeconomic factors derive from the political environment. Macroeconomic changes can result from federal legislation in the marina and boating industries and changes in regulations from the state Depart-ment of Environmental Protection and Energy. An example of such a change is the Luxury Tax Bill, passed by Congress in October 1990, specifying a 10% federal levy on luxury items such as expensive foreign cars and boats priced above $100,000. This tax bill resulted in substantial declines in the boat industry in shoreline states, significant drops in marina employment, and large downturns in boat sales. This tax, coupled with the recession of the 1990s and the savings and loan crisis, caused financial difficulties for many marina owners. Fortunately, the luxury tax was repealed in the mid-1990s and no longer affects the boating industry.

Socioeconomic Forces

Growth and Aging of the Population

As the population of the country grows, the potential market for marinas grows. In 1984, 40% of the population lived within 50 miles of the coastline. This percentage is expected to double by the twenty-first century, increasing demand over the next few decades.

The post-World War II baby boomers will add to the demand for more leisure activities as they continue to age. Significantly, boaters have a median age of 40. With increasing life spans and an improved quality of life, more of the elderly will be able to enjoy boating. This increased demand should result in diminished vacancy rates and higher slip rental pricing.

Wealth Accumulation

Fortunately, senior citizens can afford the leisure activities they demand, since almost half of the nation's personal wealth is vested in its senior citizens.

In the twenty-first century, the percentage of the U.S. population living within 50 miles of the coastline is expected to double the 1984 figure of 40%, increasing the demand for leisure boating activities.

Seniors have the most disposable income, and marina activities are classified as a luxury item. Boating is ranked third in the total number of senior participants, following golf and bowling. On the other hand, seniors have the least liquidity of all the age groups. It is likely that this group will demand smaller, more cost-effective boats.

The Family Unit
With a significant divorce rate, there are a greater number of households and fewer members in each household. Studies show that a male role model helps to develop recreational interests in children. Since there is less male influence in some one-parent households, marinas may have to take a proactive approach to attract younger clients. Services could include day care facilities, playgrounds, and boating classes.

Local Market Forces and Marina Characteristics

The local market forces and marina characteristics are more important variables to the success of a marina than the global macroeconomic environment. Submarket analysis should be a part of every appraisal, and this is usually accomplished through a slip rental survey. This survey includes details about marinas competitive with the subject such as the number and length of slips, vacancy rates during the summer and winter, the location of the waterways, services available, slip rental rates during various times of the season, and any other relevant information. The rent per lineal foot of boat, derived by dividing the annual slip rental rate by the boat length, indicates the different price submarkets in the area. The sample appraisal in Chapter 6 includes an example of a slip rental survey.

Characteristics of a marina that affect its ability to compete and stay in business include the supply of marinas, site and building characteristics, quality of management, potential for dockominium approval, dry rack storage potential, and business elements. These topics are introduced here and will be covered in greater detail in Chapters 2 and 5.

Supply of Marinas
During the early 1990s, marinas became more expensive to develop and viable land grew scarce. Thus marina construction dwindled in many shoreline states, and the expansion of marinas was also curtailed. Since there was a limited supply of marinas in the market and demand exceeded supply, slip rental rates began to increase and marinas became more profitable. As the population continues to grow and baby boomers age, slip rental rates should continue to increase.

Site and Building Characteristics
A large number of site characteristics affect the viability of a marina. The amount of upland, the amount of land under water (called the basin), the number of legally allowable slips, visibility, location, the presence or absence of utilities, the condition of the improvements, the number of improvements, the type and interaction of the operations of the buildings, and other characteristics affect the performance of a marina. Chapter 2 expands the discussion of site and building characteristics.

The amount of upland and land under water, the number of legally allowable slips, the presence or absence of utilities, the number and condition of improvements, the type and interaction of buildings, and other site characteristics affect a marina's performance.

Quality of Management

To be profitable a marina needs a manager who understands basic business practices, the labor-intensive demands of the job, and liability issues resulting from operating a marina. First of all, the marina owner or dockmaster has to be aware of the slip renter's needs and provide improvements or services to meet these needs. Laborers with specialized knowledge are required for many positions, making labor costs higher than labor costs for most commercial properties and requiring knowledgeable management of personnel.

The safety of the marina is also a critical concern of the manager. The docks must be clean to appeal to boaters, and the slips should be routinely inspected for problems that could result in injury or death. The mechanical equipment for hauling boats must be in good working order at all times. Proper safeguards are needed for the gasoline pumps and tanks, and they must continually pass municipal inspection. Since severe wind and high waves can shut down a marina, proactive damage containment is very important. These are only a few of the dockmaster's responsibilities.

Potential for Dockominium Approval

A dockominium allows the boater to have fee simple ownership of the slip water space. This ownership is similar to the ownership of air rights in a condominium. Although this form of investment is not prevalent, an appraiser should be aware of the potential for dockominium approval.

Dockominiums became a hot topic in the 1980s because many markets had excess demand for slips. Dockominium conversion (selling the slips instead of renting them) was seen as a way to capitalize on this demand. Many marina owners were excited about dockominium conversion and believed that it was a long-term trend. Some marinas were bought and sold at inflated prices because of the perceived dockominium potential which was never realized.[2]

To complicate matters, some assessors overinflated their value estimates, and many dockominiums carried excessively high real estate tax burdens as a result of this action. Speculative conversion resulted in excess supply in some markets, eliminating the viability of this investment. Since real estate patterns repeat themselves from time to time, appraisers should be aware of the history of dockominium conversion.

---

2. Dockominium conversion marina sales plagued by excessively high purchase prices should not be used within the appraiser's sales comparison analysis.

Dry Rack Storage Potential

Dry rack storage allows the vertical warehousing of boats on metal racks. The potential of dry rack storage can affect the value of the marina and its highest and best use. Dry rack storage might be used when land costs are at a premium and when it is cost- and time-prohibitive to construct marinas. It is possible to generate more revenue with dry rack storage.

Business Elements

The addition of interrelated businesses that will work in conjunction with the marina businesses can change the highest and best use of the marina and its value. Another item to consider is equipment specific to marina operations. These topics are covered in Chapter 5.

Legal and Environmental Issues

The legal and environmental issues that affect marinas are diverse and vary from state to state. Table 1.1 provides the telephone numbers of the regional Environmental Protection Agency offices that oversee most of the legal and environmental issues for the states. The regional EPA offices can refer appraisers to other state agencies that handle marina concerns within a particular jurisdiction.

The responses in Figure 1.1 were taken from a 1996 study conducted by the International Marina Institute.

TABLE 1.1 REGIONAL EPA OFFICES WITH JURISDICTION OVER MARINAS

State	Agency Name	Telephone Number
Connecticut, Maine, Massachusetts, Vermont, Rhode Island and New Hampshire	EPA Region 1	(617) 565-3402
New York, New Jersey, Puerto Rico and the Virgin Islands	EPA Region 2	(212) 637-5000
Delaware, District of Columbia, Maryland, Pennsylvania, Virginia and West Virginia	EPA Region 3	(215) 566-5000
Kentucky, Tennessee, Georgia, North Carolina, South Carolina, Mississippi, Alabama and Florida	EPA Region 4	(404) 562-9900
Illinois, Indiana, Michigan, Minnesota, Ohio and Wisconsin	EPA Region 5	(312) 353-2000
Arkansas, Louisiana, New Mexico, Oklahoma and Texas	EPA Region 6	(214) 665-2200
Iowa, Kansas, Missouri and Nebraska	EPA Region 7	(800) 223-0425 (913) 551-7003
Colorado, Montana, North Dakota, South Dakota, Utah and Wyoming	EPA Region 8	(800) 227-8917 (303) 312-6312
Hawaii, California, Nevada and Arizona	EPA Region 9	(415) 744-1500
Washington, Oregon, Alaska and Idaho	EPA Region 10	(800) 424-4EPA (206) 553-1200

FIGURE 1.1 BIGGEST OBSTACLES TO GROWTH

What follows are the individual responses to the question, *"What are the biggest obstacles to growth for your marina?"*

- No waterfront available, lack of funding
- Government/Federal, state and local regulations
- Size of property—no room for growth
- Available uplands—no room for growth
- Government agencies
- Permitting and local conservation
- Corps of Engineers
- County permitting
- Land restrictions
- No additional water space for lease
- Money—lack of funding
- State approval
- Lack of water space and lack of land for improvements, costs to develop
- No space for growth
- Permitting
- Government regulation
- Marina is landlocked, no room for expansion
- No land
- Not enough room with deep water
- Town code, restore 20% per year with no configuration changes
- Lease, do not own
- Footprint is built to capacity
- Moderately easy on land side; difficult on water side
- Space, financing, regulations
- Cost of adjacent property
- Land—difficult using the wetland or our property
- Channel constraints
- Obtaining permission from U.S. Army Corps of Engineers
- Neighbors
- Local ordinances limiting growth
- Limited space on land—"built at" over water and nowhere to go
- Cost of dredging, limestone rock bed, flooding, TVA regulations, Ghant restrictions, Corps of Engineers restrictions
- Permits/Financing
- Physical barriers and environmental regulations
- Lack of property, water surface and depth

(continued)

FIGURE 1.1 BIGGEST OBSTACLES TO GROWTH (CONTINUED)

- Shallow water
- Local ordinance and permit requirements
- Limited land and water area
- Fixed harbor lines, city zoning, lack of property
- State government
- Bureaucracy
- Real estate on waterfront not available or prohibitively expensive
- Space to expand was planned
- Space (land and water) and zoning
- Limited waterfront dockage
- No room in harbor; land assessment
- Getting approval from the Department of Natural Resources
- Lack of land
- DNR/Corps of Engineers/City
- Permits and loans
- Difficult to expand dockage; moderately easy to expand uplands; shoreline regulations
- No more room
- PA Department of Environmental Resources
- Permits, lease options
- Supply and demand
- No area available
- Land not available
- Corps of Engineers
- Local power authority, federal government
- Space
- Location set in Dade County Park
- Permitting and capital
- 1. Demands for slips and commercial space 2. Raise capital 3. Environmental
- 1. Expansion would have to be off-site
- 1. Government
- 1. Money
- 1. Permits 2. Available space
- 1. Permitting
- 1. Permit process—Corps of Engineers
- 1. Expense
- 1. Non-conforming use—70 residents within 200 feet of property line

Courtesy of the International Marina Institute, *Financial & Operational Benchmark Study for Marina Operators* (Wickford, Rhode Island: International Marina Institute, 1996). Equivalent information for the top 25% of marinas participating in the study are also available from this publication.

SUMMARY

This chapter presents an overview of marinas including definitions and descriptions, a brief investment history, and a description of the factors that affect marinas. Marinas can be classified as recreational marinas, yacht clubs, or boatyards. Recreational marinas cater to boaters who use their boats primarily for pleasure. Yacht clubs provide more services than recreational facilities and may be expensive to join, and boatyards are primarily commercial facilities, offering on-site services such as boat dealerships and repair yards.

Marinas differ from most commercial properties in several ways. The value of a marina's land and site improvements outweighs the value of improvements in most commercial projects. Other differences include restrictive zoning and fluctuating income for marinas, quality of materials used for construction, location within flood zones, and specialized management costs needed to operate a marina.

During most of the 1990s, marinas were perceived as a poor investment, but this situation seems to have reversed itself. However, the specialized nature of marinas remains, and capitalization rates can have a wide range. To discover the appropriate capitalization or discount rates for a subject, an appraiser needs to speak with participants in the market.

Marinas are subject to the same macroeconomic and microeconomic variables that affect other properties. These include the political environment, socioeconomic forces, local market forces and marina characteristics, and legal and environmental issues. Each of these factors will be discussed in further detail in the chapters that follow.

CHAPTER TWO

Site and Building Description and Business Elements

In many ways, appraising a marina site is similar to appraising any other piece of property. Topography, frontage, size, shape, and other characteristics are easy to describe. Most marina buildings are used for obvious purposes, and these buildings and their purposes can be explained easily. The same is true of business elements. Both marina sites and commercial properties often benefit from complementary businesses that are located near them. However, in spite of the similarities between marina properties and other commercial properties, there are also specific items that differentiate marina properties from other properties. This chapter will describe site and building characteristics and business elements that are unique to marinas and some of the factors that an appraiser needs to consider when appraising a marina.

SITE DESCRIPTION

Some of the site-specific items that are particular to marinas include location factors, riparian rights, protection from waves, a wide variety of on-site utilities and services, water quality, water frontage and depth, breakwaters, flood zone classification, wetlands and marshes, bulkheads, piers and slips, docks, retaining walls, slip length, dredging, deepwater status, parking, and dry rack storage, etc. These items are described in this chapter.

Location Factors

Many location factors influence the success of a marina. Some of the more important factors include proximity to population centers; convenience to waterways and outlets; visibility from commercial roadways; amount of upland available for boat, trailer, and automobile parking; potential for townhouse or condominium development; proximity to large fishing areas; and water depth. The types of services and businesses offered by a marina and the interrelatedness of them are directly dependent upon these and other location factors. The appraiser has a duty to determine which factors are most important and to analyze them accordingly.

Riparian Rights

Riparian rights and riparian grants are similar. A riparian right is defined as

> The incidental right of an owner of land abutting a body of water
> to use the water area for piers, boat houses, fishing, boating,

navigation, and the right of access for such purposes, limited by public need if on a navigable stream.[1]

With a riparian grant, the owner purchases the land underneath the water, with certain restrictions. Most of the seaboard states sold their inventory of riparian grants decades ago, and these grants are rarely offered for sale today. No payments need to be made on riparian grants, but annual payments must be made on riparian leases.

There are rarely differences in value between riparian leases and grants, but it is important for the appraiser to determine if the riparian water rights are leased or granted. In addition, the appraiser should verify that the payments are current and that the riparian rights have not expired. If a riparian rights lease has expired, the marina may have to close. This will have a significant impact on value.

Protection from Storms, Waves, Wind, and Ice Damage

Marinas are more susceptible to damage from the elements than commercial properties are. To reduce damage from ocean storms and waves, marinas are usually located in some type of protected cove or inlet. Man-made protected coves may be created for marinas located on the ocean. Marinas located on freshwater bays are less likely to experience wave damage; however, damage from ice is a distinct possibility. To analyze potential ice damage, the appraiser should research the history of the marina to see if there is a history of icing. If so, the appraiser can determine the degree of damage that occurred to the boats in wet storage or marina slips in the past.

Utilities

To be competitive with other marinas, marina properties must provide more utilities than most commercial properties provide. With commercial properties, it is often sufficient to have municipal water and sewer available to the buildings. With marina properties, it is necessary to have one or more of the following utilities present at the slips:

- Water for houseboats
- Sewer for houseboats
- Electricity, allowing boaters to do minor repairs on their boats

---

1. Appraisal Institute, *The Dictionary of Real Estate Appraisal,* 3d ed. (Chicago: Appraisal Institute, 1993), 312.

- Bottled gas
- Gasoline for boat motors
- Telephones for houseboats
- Cable television for houseboats
- Sanitary sewage and waste removal

Marinas providing water, electrical, and sewer utilities are classified as full-service marinas. Those lacking one or more of these utilities are often at a disadvantage. This disadvantage might be reflected in the occupancy rates during the spring and summer seasons and the rates that can be charged for slip rentals.

Gasoline pumps may be an advantage or a disadvantage. The appraiser should carefully note how the gasoline is stored, the type of holding tank, the capacity of each tank, and the last testing date or replacement date of the tank. Older, buried steel gasoline tanks may leak, sometimes unnoticed, into the bay. Leaking tanks can create serious environmental problems. Many states have enacted laws requiring the removal of steel gasoline tanks when a property is sold.

On-Site Services

By providing popular services, marinas receive additional revenue and prevent customers from looking at other facilities for the services they need. This is also true of many commercial properties that must provide services to compete with other properties.

The most common service offered by a marina is boat storage. Boat storage may be provided by wet slips, open air ground storage, or dry rack storage. Another service that is in high demand is boat repair and washing. The larger the boat, the greater the need for repair. Some repair buildings have one or more channels, allowing boats access into the building where they are moored and repaired. Winterizing service, another popular service, is related to repair service. As the name implies, winterizing prepares a boat for the coming cold weather season. Workers correct minor damage and wear from the summer season. During the winterizing process, repairmen may notice work that needs to be done on the boat and suggest the repair service.

Full-service marinas provide gasoline pumps, water hoses, and fire suppression systems, as shown here, along with electrical and sewer utilities and other services.

Water Quality and Fishing Grounds

The quality of the water and a marina's proximity to recreational fishing can affect the number of customers that a marina has. Rivers that have received public attention as a waste-dumping area may be seen as less desirable places for boating. Customers interested in recreational fishing may choose one marina instead of another because of the marina's location near fish, crab, or lobster grounds. The appraiser can check with the state's wildlife agency to discover proximity to fishing grounds and the water quality commission or agency to check the water quality.

Water Frontage, Water Depth, and Land Area

The boat slip count is analogous to a room count in a hotel and has a significant bearing on value. If all other factors are equal, a marina with more slip counts than a competitor has a greater opportunity to increase income than the competitor has. The amount of water frontage and the ownership of the land under the water basin control the number of allowable boat slips. With an oceanfront marina, the typical minimum depth at dockside should be six to seven feet.

The amount of upland is also significant. It must provide parking spaces in the boating season and dry boat storage in the off season. A large land area will allow future expansion with ancillary uses.

Breakwaters

Unlike most commercial properties, marinas can be threatened by strong water currents and waves. To protect buildings and slips, a breakwater is usually constructed. A breakwater is a barrier or structure that stops or slows water currents and waves. Breakwaters can consist of a rocky mound, a floating breakwater structure, or a solid wall breakwater that is affixed to select piers.

Sometimes breakwaters cause problems. They can reduce the natural flushing of the basin or impede the flow of water, allowing additional sedimentation buildup. For marinas located in northern climates, ice can cause the same kind of problems.

In extreme cases, major storms have damaged the piers that have solid breakwaters attached. The appraiser should be aware of the cost involved to repair or replace a breakwater damaged by a storm. This cost can be reduced by having multiple marina owners contribute to a common breakwater fund.

Flood Zone

Commercial property builders try to avoid flood zones, but marinas are usually located within the most severe flood zones. The appraiser of a marina site will frequently find that buildings used for marina operations have flooded on numerous occasions.

If home development is contemplated, many jurisdictions require buildings in flood zones to be elevated on piles to minimize flooding. The appraiser must carefully determine the exposure of the marina site to flooding and the potential ramifications if flooding were to occur.

Wetlands and Marshes

The owners of marina properties must be aware of environmental regulations and laws, just as the owners of commercial properties are. However, marina owners are more likely to encounter problems with environmental regulations because wetlands and marshes are commonly found on or along the site of a marina. To prevent disruption of the ecosystem, federal law provides jurisdiction of the wetlands to the U.S. Army Corps of Engineers. In addition, many states have enacted legislation that further defines what can be done on or around wetland areas. Some wetland areas may be the home to one or more endangered species.

In a 1996 survey conducted by the International Marina Institute, respondents listed using wetlands and obtaining permission from the Army Corps of Engineers as obstacles to growth. (See Figure 1.1 at the end of Chapter 1.) Environmental regulations have halted much potential marina development. The appraiser should be aware that expansion of a marina may be impossible if the marina is near a wetland.

Expenses to Repair Bulkheads, Piers, Walkways, and Slips

Some of the site-specific items of a marina are very expensive to repair. These items include the bulkhead, piers, walkways, and slips. The condition of these items can affect value, slip rental rates, and the safety of patrons.

A bulkhead is defined as follows:

> A retaining wall that is backed with solid fill and erected along the water to extend the upland out to the bulkhead line; serves as protection against tidal or watercourse erosion of land.[2]

A rotted or damaged bulkhead is an expensive item to repair and has a damaging effect on value. Repairing or replacing a bulkhead requires permission from one or more regulatory agencies due to potential disturbance of the sediment. It can take a long time, sometimes years, to complete all the paperwork necessary for this process.

Slips are defined as the water space between two piers. The width and length of the slips, as well as the minimum water draft (the depth of the water measured at low tide), will determine the size of the boats that can be moored at the marina. The appraiser must adequately describe the slips since the number and condition of slips directly affect slip rental rates and the value of the marina.

Wooden piers and walkways can present a safety hazard if they are in poor condition. The appraiser should be aware of the safety and value issues of all these items.

Docks

Marinas usually have floating or fixed docks. Tides, the size of the boats, and the weight of transported loads determine the kind of dock found at a marina.

Fixed docks are preferred in areas where tides are nominal or nonexistent and the boats are quite large (generally above 50 feet in length). Also, if the

---

2. Ibid., 42.

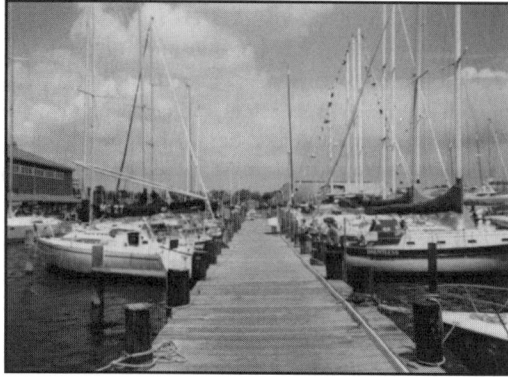

Fixed docks and piers like this one are preferable in areas with small tidal variation and for larger boats (above 50 feet in length).

weight of the load being transported over the dock is significant, such as fish from a commercial fishing operation, the marina should have a fixed dock. Fixed docks come in a variety of designs with different materials, and they are usually more expensive than floating docks. Piles provide support for the fixed dock. Piles are usually wood that has been specially treated to resist the corrosive effects of saltwater. Concrete or steel can be used for piles, but the cost of these materials usually outweighs any advantage of using them. Piles can be fixed into the sediment or rock.

Marinas located in areas with significant differences between low and high tides usually have floating docks with staircase ingress and egress. There are several basic types of floating dock frames: wood, concrete, or metal. All of the ancillary items such as utilities, flotation, decking, protection cleats, connectors, and anchorage attach to the frame. Wooden floating docks are designed in a lattice structure to provide additional strength with flotation devices underneath and wood decking above. Concrete floating docks consist of a polystyrene foam block that has concrete along the top and sides and sometimes the bottom. Wooden planking is attached to the top, and the various docks are connected with heavy timber.

Metal floating docks can take on a wide variety of forms including large steel tanks filled with air, a steel pipe lattice structure with air ballast, steel or aluminum with pontoons underneath, and steel and aluminum trusses with some type of flotation system affixed (the most common floating dock). Floatation devices include metal drums, hollow or filled metal pipes, Styrofoam billets, fiberglass tubing, and hollow or filled polyethylene tubs or concrete boxes. Fill material can include polystyrene, polyurethane, or Styrofoam. Decking can include wood, concrete, fiberglass, steel, and aluminum.

It is important for the appraiser to determine the appropriate type of dock. If the wrong type of dock is in place and the market recognizes a difference, the appraiser may need to make adjustments.

Retaining Walls

Sheet pile retaining walls are designed to prevent waves from eroding the site. When they fail, an expensive repair and fill operation must be undertaken. There are two types of retaining walls: cantilever and anchor. Cantilever walls consist of steel sheets that are driven into the sediment to a depth sufficient to resist the deflection of the waves and the erosion of the soil. These walls are seldom used because they are more expensive, relatively unanchored, and can shift under significant stress. Anchored walls, the preferred type of walls, provide a number of support points that will counterbalance the wave force generated against them.

Retaining walls are usually constructed of wood since wood is economical to use. The walls should be constructed of concrete, steel, or aluminum if boats bring in abnormally heavy loads or loads that require higher than normal dock height.

Slip Length and Turning Radius

Functional obsolescence exists if boat slips are used for boats smaller than the size the marina accommodates. This type of obsolescence also exists if the ingress and egress are impaired. An example of functional obsolescence is a marina that caters to large boats and decides to accommodate small boats in the larger slips because the market demand is for small pleasure craft under 25 feet in length. This results in a higher vacancy rate and lower revenues. Another example is a marina with slips that are too narrow for the boats to turn or an aisle width between two piers that does not provide enough room for turning. This results in an impaired ingress and egress. (A similar situation exists at industrial facilities with a turning radius that is too narrow for trucks.)

The appraiser must be aware of these situations when evaluating a marina. It may be necessary for the appraiser to make a functional obsolescence adjustment.

Dredging

Dredging is the solution to the buildup of siltation to an unacceptable level. Siltation builds up as currents flow in and out and carry sediment to an area where there is relatively little water movement. The sediment loses velocity and sinks in these areas. As sediment builds up, the slips become more shallow. Most pleasure boats need a minimum of one to two feet of water below their propellers during low tide, referred to as a shallow water draft. Six to seven feet at mean low water will accommodate virtually any of the large yachts or sailboats.

Most marina owners maintenance dredge periodically to prevent excessive buildup of siltation. Unfortunately, there are problems associated with dredging. These include excessive cost, long time frames for obtaining the multiple permits necessary to dredge, lack of places that accept the dredged material, and the possibility that pollutants and contaminants are embedded in the dredged materials.

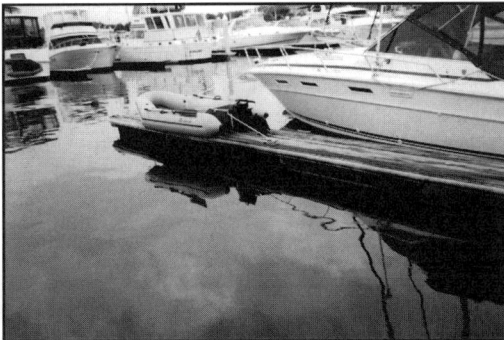

Floating docks like this one better accommodate large differences between low and high tides and are generally less expensive to build than fixed docks.

The cost to obtain the permits and approvals can exceed the expense of the dredging itself, and it can take several years for the permits to be issued. Since disturbing the beds can disrupt the local ecosystem, the Department of Environmental Protection and similar state agencies have enacted laws in many states that complicate the approval process. Major dredging or development projects may require expensive, time-consuming environmental impact studies. To minimize the paperwork process, many marina owners cooperate and submit dredging proposals together. Timing the dredging to occur during the off-season prevents disruption of services and revenues.

In addition to environmental impact studies, marina owners may need to have the sediment tested for contamination. Although strict laws effectively prohibit dumping, past practices may have resulted in some sediment contamination. Some types of sediment, such as materials with high organic silt and clay contents, are prone to collecting pollutants. Contaminated materials can still be removed, but the number of sites that will accept these materials are few. Creative solutions include using the material as fill underneath roadway projects, creating man-made islands, depositing the material in solid waste landfills, using the material for dikes or marshes, or dispersing it in areas where there is no sediment pollution. Incineration may also be possible.

To determine areas where siltation builds up, the appraiser needs to be aware of the movement of the currents within the bay or river and the degree to which siltation occurs. The U.S. Coast Guard has mapped every bay and tidal waterway, and these maps are a prime source for determining water movements. Marinas with rapid siltation buildup usually have less marketability due to the cost of dredging and the resulting downtime. This can translate into a lower value. Similarly, bodies of water with a history of sediment contamination may have less marketability and value.

Deepwater Slips

Deepwater slips, those that can accommodate 40-foot or longer boats, can be particularly valuable. There is usually more demand for deepwater slips than supply, and this is reflected in the rental rates. A marina that has deepwater slips and no pleasure boat slips is often used as a commercial fishery operation or yacht harbor. Deepwater slips add diversity to an operation because yachts and tourist boats can be accommodated, often at premium rates.

In highly urbanized areas or waterfront districts, there may be problems associated with deepwater slips. Sometimes underground rights have been sold to allow underground cables, subway systems, or pipelines. These items may impact the replacement of deteriorated piles or may even result in a different dock arrangement, such as a floating dock in lieu of a fixed dock. If the subject is a marine terminal with deepwater slips, it may suffer from external obsolescence from nearby deteriorated buildings or an inadequate roadway or support system. Buildings that extend onto a pier are usually too expensive to replace. Dumping practices in urban areas are another consideration. The beds may be so polluted that dredging would be prohibitively expensive, or sedimentation may have built up under the open piers requiring removal of the building or pier to redredge.

Parking

Parking at commercial properties and at marinas becomes a problem if there are not enough parking spaces. During summer weekends, especially the Fourth of July weekend, marina parking needs may compete with open boat storage, the efficient movement of boats, the movement of boat handling equipment and marina vehicles, and the required parking for other on-site businesses such as restaurants.

There has been considerable debate about the number of parking spaces necessary for the proper operation of a marina. Part of the confusion resulted from local planning boards assigning marina parking ratios similar to those of retail or office space ratios. Planning officials did not realize that boating is seldom a solitary recreation, and most boaters carpool to a facility.

Early studies revealed a parking ratio ranging from 1.5 to 2.5 cars per slip. Two cars per slip became the rule-of-thumb ratio. However, numerous studies conducted during the 1970s and 1980s indicated that this estimate was overinflated. The most recent rule-of-thumb ratio is one car for every two boat slips.[3] If the subject's parking ratio is significantly less than the recent figures, this may result in a backlog during the peak summer holidays. If the ratio is significantly higher than the recent figures, there may be enough land for additional development or building expansion. (This should be considered within the appraiser's highest and best use analysis.) Many marinas have too many parking spaces because of the overinflated estimates of planning boards.

Dry Rack Storage

Dry racks are metal racks that are used to warehouse boats. They are usually two or three levels high. Heavy boat lifts are used to move the boats to the higher elevations. Dry racks can be a good source of revenue because they allow additional vertical storage on the site, especially when land is at a premium. They also allow marinas to store boats on-site during the winter when few or no boats would normally be kept on the water. The potential for dry rack storage should be considered in the highest and best use section, depending on the amount of land and the probable location of dry rack storage on the site.[4]

BUILDING DESCRIPTION

As mentioned in the introduction to this chapter, marinas are similar to and different from commercial properties not only in site-specific ways, but also in the description of buildings.

Just like commercial properties, marinas can have mixed-use buildings. There may be mixed-use buildings with residential or office components, boat sale and repair buildings, one or more office buildings, bait and tackle shops with commercial fishery buildings, and others. In addition, most marinas have a crane or lift to lift boats in or out of the water. These improvements can be owner-occupied or leased. The appraiser should determine the ownership interest of each building and include such information in the report.

Unlike many commercial buildings, marina office buildings are usually constructed of low-quality materials and are basic in design. Since marina buildings are located in a flood zone, it is more cost-effective to construct inexpensive buildings. The appraiser may feel that elaborate buildings are

---

3. Neil W. Ross, *Auto Parking in Marinas* (Wickford, Rhode Island: International Marina Institute, 1989), 1-23. This study indicated that the number of cars per boat slip ranged from 0.21 during the August weekday low use period to 0.71 during the Fourth of July holiday weekend peak.

4. For more information, see John A. Simpson, "Appraising Marina Proposed Dry Rack Storage," *The Appraisal Journal* (July 1998).

Although travel lifts are personalty, they are considered necessary to a marina's operation and are included in a property's value like a hotel's personal property.

superadequate because of the potential of flood damage. Marina office buildings often show interior signs of water damage from flooding, and the appraiser should carefully note this damage. Some office buildings have one or more residential units above grade that are used by the manager or a tenant.

Improvements on commercial properties often have alternatives uses. Many of the improvements at a marina site may not have an alternative use, because a marina attracts a specific type of clientele. For instance, an on-site restaurant will usually cater to marina patrons and the immediate surrounding area. Without the support of the marina, it may not be financially feasible for the restaurant to stay in business. The appraiser should be aware of these conditions, especially when the marina is not performing well or is insolvent.

Bait and tackle shops are another business that has little value or utility if the marina is inoperative. These shops are usually small, shack-like, wooden improvements located on or near the slips. Sometimes they are located on the site itself. No special equipment or design is present; the appraiser can easily describe these buildings.

Repair buildings on marina property can range from small, wooden, shed-like structures (similar to bait and tackle shops) to large, metal, warehouse-like facilities. There may be fixtures or equipment specific to the repair of one class of vessel. If so, this information must be considered and stated within the appraisal. Some buildings may have a slip that allows a boat to moor within the building itself, thus saving a time-consuming crane moving process. However, many repair operations will have cranes to lift boats in and out of the water. When an overhead crane is attached to the building, it is a fixture and is valued as part of the appraisal. When the crane is mobile (usually referred to as a travel lift), it is personal property and is usually valued with the marina. The appraiser must be aware that these travel lifts are present in the facility and state this fact in the report. Even though lifts are personalty, they are considered necessary to the marina's operation and must be included in the property's value much like a hotel's personal property.

A lack of travel lifts can adversely affect value because many boat owners will be unable to get their boats into the water. They may choose not to rent a slip at the marina for this reason. Only owners with small pleasure crafts will be interested in renting because they can float their boats on and off of trailers by way of the submerging loading ramp. It is rare for a marina not to have a crane or lift.

BUSINESS ELEMENTS

In addition to the improvements and fixtures on the site, the appraiser must be aware of the interrelated businesses and management skill inherent in larger marina operations. These businesses can include boat sale operations, repair operations, restaurants, motels, and condominium sites associated with the marina sales. Marina sites that are well planned include businesses that complement each other, drawing customers that may purchase from a variety of the businesses. The examples below explain how interrelated businesses and management personnel affect the profitability of the marina.

Example 1

A boat sale operation with marina slips will realize higher profits than one without slips because a boat purchaser may be able to obtain a slip for several months until he or she finds a more permanent rental situation. When the slip rental market is tight, providing slips is especially important for boat sale operations because most potential boat purchasers will not buy a boat without a slip to house it. Boat or motor franchises that provide slips include the Bayliner Boat Line, the Wellcraft Boat Line, the Mercury Motor Line, and the Johnson Motor Line, to name a few. These operations have an intrinsic advantage over facilities without slips, especially when slips in the area are fully leased.

Example 2

The presence of a major boat line usually increases the repair profits of the marina. Some marinas have a franchise agreement with a large boat sale/repair operation such as a Sea Ray franchise. This type of franchise usually commands a larger market share of boat sales and repairs than a facility without a brand name. The repair operations alone can increase demand for slips due to the convenience of having a repair facility on-site. The presence of the franchise can result in increased profits for all operations.

Example 3

Some marina sites have been developed in conjunction with condominiums or townhouses. This type of development creates a ready-made market for marina slips because many purchasers want a slip. Most new marina sites in the 1990s have been developed with these types of real estate concepts.

Example 4

A restaurant and motel or hotel may be part of the marina site and help to increase the profits for the other businesses at the site. Restaurants usually require a highly visible commercial location. Most marina restaurants will partially depend upon marina clientele for support. These restaurants normally specialize in seafood which fits well with the nature of the area. Motels or

A boat sales and leasing operation will lose profits if it does not have slips of its own to house boats for customers until a permanent slip is found.

hotels usually receive a relatively small percentage of their income from a marina; their competitive environment should exceed the trade radius of the marina. As a result, they may not provide much additional revenue for the other business elements of the marina.

The more utilities and services offered, the greater the need for a manager with specialized knowledge and management skills. A marina manager must be able to handle the specific needs of boaters as well as operate a business. Like most industries, repeat business is important to the success of a marina, and the manager must be able to anticipate the needs of customers and provide the needed services. To determine if a facility is meeting the needs of the renter, appraisers can envision themselves as customers and ask the following questions:

- Does the facility cater to active recreational boaters who like to fish or pleasure boaters who like to sail?
- Is this facility merely a warehousing operation for boats?
- Are there adequate utilities?

If a marina site lacks competent management, the appraiser should include this fact in the report. Competent management is an important element that helps to determine the profitability of the marina.

SUMMARY

There are similarities between marina properties and commercial properties, but there are also specific items that make marina properties different from commercial properties. An appraiser must be aware of the unique site and building characteristics and building elements of marinas.

Some of the site-specific items described in this chapter include location factors, riparian rights, protection from waves, a wide variety of on-site utilities and services, water quality, water frontage and depth, breakwaters, flood zone classification, wetlands and marshes, bulkheads, piers and slips, docks, retaining walls, slip length, dredging, deepwater status, parking, dry rack storage, etc.

There are also specific items that make marina buildings unique from commercial buildings. Since marina buildings are located in a flood zone, they are usually constructed of lower quality materials and are more basic in design than commercial buildings. Marina buildings may have mixed uses, such as residential with office, bait and tackle shops, boat sale and repair operations, and commercial fishery buildings with bait and tackle shops. In addition, most marinas have a crane or lift to lift boats in or out of the water. Unlike commercial properties, many of the improvements at a marina site may not have an alternative use. Without the support of the marinas, businesses such as restaurants, bait and tackle shops, boat repair operations, etc., may go out of business.

Management and the interrelationship of businesses make up the business elements of the marina. The management of a marina requires specialized knowledge and skill. The manager must be able to anticipate the needs of customers and provide needed services. Marina sites that are well planned include businesses and services that complement each other, drawing customers that purchase from a variety of the enterprises. The presence of a major boat line, franchise agreements with boat sale/repair operations, etc., can increase the profitability of the marina. Appraisers should include an evaluation of management and businesses in the appraisal report.

CHAPTER THREE

Highest and Best Use

To build on the material from Chapter 2, it is necessary to review the major points. The use of a property as a marina site is a specialized use. Managing a marina requires specialized knowledge. Marinas have a wide variety of specific site and building characteristics. Since marinas are located on the water, they are subject to zoning restrictions affecting other uses such as industrial and commercial activities.

Analyzing the highest and best use of a marina site requires detailed study of these and other factors mentioned in Chapter 2. This analysis requires the use of the four traditional tests used to analyze all properties. These tests and their applications to marinas are described in this chapter.

HIGHEST AND BEST USE AND THE FOUR TESTS

Highest and best use is defined as follows:

> The reasonably probable and legal use of vacant land or an improved property, which is physically possible, appropriately supported, financially feasible, and that results in the highest value.[1]

In the case of vacant land, the analysis of highest and best use is performed for the land only. In the case of improved properties, it is performed separately for the land and the buildings. For both types of properties, a use must pass four consecutive tests:

1. Physically possible—the use could be physically accommodated at the site.
2. Legally permissible—the use is legally allowed for the property.
3. Financially feasible—the use will produce a positive return.
4. Maximally productive—the use is the one that produces the highest return.

FACTORS TO CONSIDER IN DETERMINING HIGHEST AND BEST USE

There are factors specific to marinas that must be considered when evaluating the highest and best use of a marina. These include condition of the bulkhead,

---

1. Appraisal Institute, *The Dictionary of Real Estate Appraisal,* 3d ed. (Chicago: Appraisal Institute, 1993), 171.

marina, and buildings; potential for dockominium approval; compatible uses that could be accommodated at the facility; potential for townhouse or condominium development; and allowable uses under the zoning.

As discussed in the previous chapter, business elements can have a marked effect on profitability and highest and best use. The addition of a new boat sale enterprise in conjunction with an existing repair operation and marina slips can result in increased profitability for each area of the marina. Sometimes the addition of one or more boat and motor lines to existing franchises will generate more revenue for the marina.

Unfortunately, it is very difficult for an appraiser to isolate the business elements that may be present at a marina. There is an undefined area where the value of the improvements and their contributory value to the business enterprise can differ. In some cases, the appraiser may not be qualified to undertake the assignment. If so, the appraiser needs to follow the requirements under the Competency Provision of the Uniform Standards of Professional Appraisal Practice. These are as follows:

1. disclose the lack of knowledge and/or experience to the client before accepting the assignment; and

2. take all steps necessary or appropriate to complete the assignment competently; and

3. describe the lack of knowledge and/or experience and the steps taken to complete the assignment competently in the report.[2]

DETERMINING THE MARKET POSITION OF A MARINA

An easy way to determine the market position of a marina is to observe the typical user of a slip and the types of businesses and services offered to the typical customer. Is the customer a recreational boater who simply wants to sail or a recreational fisherman who is only interested in the sport of fishing? Is the facility near any common fish schooling areas?

A restaurant might be a viable business at a marina that serves recreational boaters, but it would probably not do well at a marina that serves slip custom- ers who are commercial fishermen. Similarly, a bait and tackle shop is important to a fisherman but means little to a recreational boater. Some types of busi- nesses, such as a marina boat store, can successfully serve both recreational and commercial boaters. These stores are often referred to as chandleries if they primarily serve commercial users.

Highest and Best Use as Vacant

Appraisers vary in the treatment of the subject marina as vacant. Some assume that there are no buildings on the site but the marina bulkhead and docks exist. Others assume that the buildings, bulkhead, docks, and site improvements do *not* exist. Since many governmental agencies will not permit bulkheads and

---

2. *Uniform Standards of Professional Appraisal Practice* (Washington, D.C.: The Appraisal Foundation, 1998), 5.

piles to be installed because of possible disruptions to the ecosystem, the highest and best use as vacant should include bulkheads and piles. (The client is also best served under this assumption.) The cost and difficulty of installing these site improvements should be stated in the appropriate places within the appraisal report.

Physically Possible

The degree to which a site may be developed depends upon its physical characteristics. The amount of upland and water frontage will determine the types of improvements that can be accommodated. Five basic questions need to be answered to determine what is physically possible on the site.

1. What marina buildings can be accommodated on the site?
2. Is sufficient open land available for normal ingress and egress of boats?
3. Is sufficient land available for on-site storage?
4. Is land available for dry rack storage?
5. What other types of separate or interrelated uses can be physically accommodated on the site?

Legally Permissible

The legally permissible uses determine how the site may be developed and its degree of development as vacant. The appraiser needs to analyze the following categories to determine what is legally permissible for the site.

Riparian rights. Are riparian rights or grants present? If not, the state may be unwilling to sell or lease any more rights, making the use of the land for a marina impossible. For simplicity, appraisers usually assume that riparian rights are in place. Appraisers also assume that the slip yield would be maximized given the size and location of the riparian rights or grants; this may result in a greater or fewer number of slips than those present at the subject. The appraiser should verify the riparian rights or grants status early in the assignment. If riparian rights or grants are not present, the appraiser may be unable to value the facility as a marina. This would alter the appraisal assignment.

Government agencies. Will government agencies allow dredging and the creation of a marina? Many states will not allow marina development due to anticipated damage to the seabed and disruptions to the local ecosystem.

Zoning. What uses will the zoning allow? Is it legally permissible to develop the site if it is located within a severe flood zone? Will the zoning allow dry rack storage, and if so, to what degree?

Does the marina require variances, and is the developer likely to receive them? Ancillary uses, such as a restaurant or hotel, may also require variances. After deciding what uses are allowable, the appraiser must consider other zoning issues such as setbacks and parking. These issues are important in determining the degree of development that will be allowed and the location of setbacks and parking.

Permits or licenses. What types of permits or licenses are necessary to operate a marina facility and its ancillary services? Will it be difficult to obtain permits to install fuel tanks? Can the necessary utility permits be obtained so utilities can be extended to the site?

Opposition. Is there a significant amount of opposition from neighbors or environmental activists which may limit development on the site? Many marinas were never approved for dry rack storage due to strong objections by neighbors and/or environmental activists.

Other uses. Are other uses, such as residential condominiums, townhouses, hotels, motels, restaurants, or dockominiums, likely to receive approval?

Obviously, the attitude of government officials and citizens can significantly impact the development of a site. To obtain an idea of the prevalent attitude, the appraiser can determine the number of marinas that have been developed or expanded in the past five to ten years. If the answer is zero, it may indicate that governmental regulations either do not allow or too severely restrict potential development. If new construction has taken place, it is likely that the various governing and regulatory authorities will permit construction, perhaps under certain conditions. Conversations with municipal and governmental officials will give a more accurate picture of the community's feelings about marina projects.

The declining number of developable sites, increased governmental regulation, and extensive environmental regulations produce an unfriendly environment for potential marina construction. In addition, developers of properties that are located on, or implicitly use, the shoreline must show that a public benefit will result from the project. As a result, most marina projects incorporate multiple uses such as residential condominiums, restaurants, and other commercial properties that coincide with a waterfront.

Most small and medium-sized developers lack the sophisticated knowledge of waterfront development needed to overcome these barriers. The lengthy, multiyear approval and development processes, combined with a lack of knowledge, create business risk and the potential for loss. Thus, high-density waterfront areas are developed almost exclusively by larger, more experienced development companies that have large amounts of capital. If the highest and best use as vacant is to develop the site with a marina (perhaps with other, related business components), there may be a limited number of qualified developers who can build the project. This fact may limit the property's development potential and adversely affect the highest and best use as vacant.

Financially Feasible

After determining the uses that can be accommodated and permitted on the site and their degree of development, the appraiser must address financial feasibility. Below are some questions that should be asked:

- Based on a slip rental and seasonal occupancy survey, what is the indication of relative supply and demand for the slips?
- If demand is materially greater than supply, is it financially feasible to construct dry rack storage?
- What ancillary services can be offered profitably? These might include boat repair and service, restaurants, hotels, motels, ship stores, bait and tackle shops, and other facilities.
- Is there enough demand to convert the slips to dockominium ownership and to sell them profitably?
- Is there enough demand for condominiums or townhouses? If so, will there be enough demand to sell them with a slip in dockominium

ownership? If there is not enough demand from potential homeowners for slip ownership, is there enough demand for slip rentals?

- Could additional utilities or services be offered, resulting in more profit?

Maximally Productive

Test 4 involves finding the most profitable financially feasible alternative use, if one exists. This use can be contrary to the highest and best use as improved due to changes in governmental regulations over the past several decades and the difficulty in constructing marina improvements on the ground level of flood zone inhibited properties.

Highest and Best Use as Improved

To determine highest and best use as improved, the appraiser analyzes the land separately from the buildings. As mentioned in the discussion of highest and best use as vacant, a use must pass the four consecutive tests.

Physically Possible

The limitations of the site may be such that no further improvements can be accommodated without compromising one or more existing uses on the site. For instance, if a condominium building were to be constructed, there might be insufficient on-site boat storage and difficulty in moving boats into and out of the water.

Legally Permissible

The appraiser needs to evaluate the following categories to determine what is legally permissible for the site.

Uses or conversions. Can the existing improvements be legally used or converted to another use? For example, can a repair building be converted to a ship store outlet?

New improvements. Would one or more new improvements be permitted? A proposed dry rack building might exceed the maximum building height requirements, and a variance might be needed. A restaurant or retail building may not be allowed under the zoning. Dredging to convert one or more pleasure boat slips to deep water slips may not be permitted.

Approvals. Will the slips receive dockominium approval?

Expansion. Can the existing slips be expanded?

Other uses. Can other uses that supplement the marina operation be accommodated under the zoning?

Financially Feasible

After determining the additional uses that are permitted, the appraiser must determine which ones are profitable. Usually, the appraiser will do a slip rental and occupancy rate survey, talk with marina and boat experts, and take a look at the market. If the addition of a boat franchise is financially feasible, this may dictate new building development or the need to upgrade existing buildings. An economic feasibility study may help the appraiser to determine if the estimated construction or investment cost bears a positive relationship to anticipated income.

ximally Productive

The maximally productive use as improved may be different from the actual use. With negotiated franchises, the difference may be large.

The legal environment surrounding marina development can have a profound effect on use. When the maximally productive use as improved is different from the maximally productive use as vacant, this is usually due to changes in the legal environment. The market may have little to do with it.

SUMMARY

Determining the highest and best use of a marina site as vacant and as improved involves consideration of the four tests used to evaluate all properties: uses must be physically possible, legally permissible, financially feasible, and maximally productive. In the case of vacant land, the analysis of highest and best use is performed for the land only. In the case of improved properties, it is performed separately for the land and the buildings.

In the analysis of a marina as vacant, some appraisers assume that there are no buildings on the site but the bulkhead and docks exist. Others assume that the buildings, bulkhead, docks, and all site improvements do *not* exist. The client is best served when the highest and best use as vacant includes bulkheads and piles.

There are factors specific to marinas that must be considered when evaluating the highest and best use. These include the condition of the bulkhead, marina, and buildings; potential for dockominium approval; compatible uses that could be accommodated at the facility; potential for townhouse or condominium development; and allowable uses under the zoning. In addition, business elements and governmental and environmental regulations can have a marked effect on highest and best use.

In some cases, an appraiser may not be qualified to undertake the assignment. If so, the appraiser needs to follow the requirements under the Competency Provision of the Uniform Standards of Professional Appraisal Practice.

CHAPTER FOUR

Valuation Process

This chapter describes the three traditional approaches to value—cost approach, sales comparison approach, and income approach—and their applicability to the valuation of marinas. The advantages and disadvantages of using each approach are presented, along with problems an appraiser might encounter with each approach.

Valuing a marina involves more than just applying the three traditional approaches to value. The business elements discussed in Chapter 2 play a part in the highest and best use and the profitability of the marina. The business value is regarded as part of the economic worth of the property. The business value is implicitly included in any value resulting from the application of the income approach. When business value is present, the appraiser should clearly state this fact in the report.

The sales comparison and income capitalization approaches are the primary two approaches used to value marinas. In most cases, the income approach (capitalization of net income) gives the most accurate indicator of value.

DATA SOURCES

To conduct a valuation and use an approach to value, an appraiser needs to collect data and analyze it. Since marinas are a specialized form of real estate, there are few data sources available to the appraiser. General data on the boating industry and marinas can be found in library sources such as newspapers; however, this type of information is of little value during the appraisal process.

Specific data can be found in several places. First, the appraiser can contact fellow practitioners or data providers for comparable sales, financial information, and verification. This is a good way to gain insight about what is happening in the local market. Second, the appraiser can contact marina owners or conduct a survey. To conduct a slip rental survey, the appraiser usually interviews marina owners. A wealth of specific submarket information can be obtained in this way. The third and best source for information is the International Marina Institute. This organization, formerly based in Wickford, Rhode Island, but now in Nokomis, Florida, has conducted a wide range of surveys of its international membership and has a large number of marina publications available. Publications of particular interest to the appraiser include the *Financial & Operational Benchmark Study for Marina Operators*,[1] the *Marina Operations Manual*,[2] and *Auto Parking in*

---

1. Courtesy of the International Marina Institute, *Financial & Operational Benchmark Study for Marina Operators* (Wickford, Rhode Island: International Marina Institute, 1996). Equivalent information for the top 25% of marinas participating in the study are also available from this publication.
2. *Marina Operations Manual* (Wickford, Rhode Island: The International Marina Institute, 1993).

rinas.[3] These publications contain important financial and operational information that appraisers may wish to consult when conducting a valuation and using one of the following three approaches to value: the cost approach, the sales comparison approach, or the income capitalization approach.[4]

COST APPROACH

Advantages of the Cost Approach

The cost approach is rarely used to value existing marina properties, but it is often included in valuing a marina to be built. The cost approach can be a very reliable valuation method for a proposed marina if it can be completed without compromising essential elements.

There are two steps an appraiser must take to do a reliable cost approach. First, the appraiser should make a detailed inventory of all the improvements on the site, including their dimensions. Second, the appraiser must accurately derive land value. This is important because a higher percentage of total value is vested in the land compared to other property types, sometimes as much as 80%.

The major cost information providers, such as Marshall & Swift, require the appraiser to know the amount of bulkheading, number of piles, amount of wood planking, and the type of boat lift system used at the marina. The appraiser will have to count the number of piles and boat lifts, as well as measure the dimensions of the planking and bulkhead. For the extraction method of valuing the land, the appraiser will need to take measurements for each comparable as well. Unfortunately, marinas receive little attention in the cost manuals (about half a page in Marshall & Swift), and even with measurements it can be difficult to apply Marshall & Swift's cost indicators with accuracy. The sample size of Marshall & Swift's estimates is small, reducing the reliability of the cost figures.

Disadvantages of the Cost Approach

As described in the previous section, the cost approach is an appropriate valuation method to use in valuing a marina to be built. The cost approach must be completed without compromising essential elements. Some of the most important reasons for *not* using a cost approach include the following elements and issues:

- It is difficult to determine land value because most heavily developed areas do not have enough marina-developed waterfront land sales to determine land value. Land value can be estimated by techniques other than sales comparison, such as extraction.

- Federal and state environmental regulations are so stringent that the subject may not be developable. If a use is not legally permissible, a primary underlying premise of the cost approach is compromised.

- Development of a vacant site may not be financially feasible given the high legal costs of obtaining all the necessary approvals.

---

3. Neil W. Ross, *Auto Parking in Marinas* (Wickford, Rhode Island: The International Marina Institute, 1989).

4. The International Marina Institute operates a Fundamental and Advanced Marina Management School which provides in-depth information and training about the operation of a marina. For additional information about the school, contact the International Marina Institute, P.O. Box 1202, Nokomis, FL 34274; phone: (941) 480-1212; fax: (941) 480-0081.

- It may take years for a project to be approved, making the cost approach of little practical use. Comparing unapproved sites to approved marina sites does not account for the time and cost of obtaining approvals.
- Further development restrictions are present when developing flood-zone encumbered land.
- The zoning may have changed since the marina was constructed, perhaps excluding marina uses.
- If the site is vacant, environmental activists and neighbors may lobby to prevent development of the site. There have been many examples of this throughout the country.
- Not every permit may be obtained to build the facility.
- If the site is vacant, there may be no capital available for marina construction.
- Estimating accrued depreciation is subjective.

SALES COMPARISON APPROACH

Applying the sales comparison approach to the valuation of a marina involves following the same general procedures used for any other property. However, a number of the specific items mentioned in Chapter 2 must be considered. These include units of measurement, upland and riparian rights, utilities and services, and buildings and conditions.

Units of Measurement

The typical unit of measurement used in pleasure boat marina appraisals is price per slip. Since most of the income is generated from slip rentals and the market purchaser uses this indicator of value, the price per slip is a logical choice for the unit of measurement. For commercial fisheries, which often have large buildings, high sale prices, and a small number of deepwater slips, the price per square foot of land or buildings may be more appropriate.

Riparian Rights and Uplands

As defined in Chapter 2, riparian rights are

> The incidental right of an owner of land abutting a body of water to use the water area for piers, boat houses, fishing, boating, navigation, and the right of access for such purposes, limited by public need if on a navigable stream.[5]

Upland is defined as the

> Land above the surface of a body of water or above a mean high water line.[6]

Riparian rights and the size of the upland affect value. Obviously, a large upland area allows for expanded parking, a greater potential for dry rack storage, building expansion, and increased potential yield if townhouses or

---

5. Appraisal Institute, *The Dictionary of Real Estate Appraisal*, 3d ed. (Chicago: Appraisal Institute, 1993), 312.
6. Ibid., 381.

condominiums are to be built. If riparian rights are present, they should be active for both the subject and the comparable sales. A facility without riparian rights cannot be operated as a marina.

Utilities and Services

The number and type of utilities and services directly affect the market position of a marina. Marinas offering electricity, water, and sewer are classified as full-service marinas. Some markets recognize this distinction and allow for a difference in value if a marina does not provide full services. To determine if a value differential exists, the appraiser must ask about local practices.

Some marinas offer related services that complement each other. This can lead to an increase in income and value. The appraiser should adjust for this increase within the comparison.

Buildings and Condition

The types of buildings and their condition have a significant impact on value. The appraiser must consider the value from structures such as boat showrooms, restaurants, motels, and hotels. Franchises and their specialized building designs must also be considered in the valuation process.

INCOME CAPITALIZATION APPROACH

For an operating marina, the income capitalization approach is the preferred approach to use. Marinas are not owner-occupied properties, and the purchaser of a marina is interested in the income it generates. The income approach is usually given primary weight when an existing income stream exists or if there is a prospect for potential income in the case of a marina being developed or enlarged.

To complete the income capitalization approach, the appraiser must obtain information about the income and expenses for each area of the marinas.

Sources of Income

Sources of income at a marina can include summer slip rentals, winter wet storage, overnight and temporary dockage, on-site storage, dry rack storage, boat washing, boat repairs, launching fees, boat services, gasoline and oil sales, and boat sales. Additional income may come from other businesses such as restaurants and stores, bait and tackle shops, and buildings such as motels, hotels, condominiums, or townhouses.

The boat length is the measure used for estimating the slip rental rates. Marinas quote slip and rack storage as a cost per lineal foot measured from bow to stern. They limit the term of usage to the season. In the northeast, it is most often from April 15 to October 15. After the term, the boat owner removes the boat or makes arrangements for off-season storage. For a boat stored on the marina land, the rate is expressed per square foot of boat, measured from bow to stern times the maximum beam (width).

Expenses

Common expenses at a marina include real estate taxes, riparian rights and grants, licenses, insurance, utilities, advertising, management, legal and professional fees, labor, repairs and maintenance, reserves for replacement, and other expenses.

Real Estate Taxes

Real estate taxes are payable on the land, improvements, and usually any riparian grants. Many marinas are overassessed because the tax assessing officials considered the land at its highest and best use. In high density areas, this use may be for development with a condominium or townhouse project and not a marina. In some cases, the value of the land is so high that the resulting real estate taxes significantly diminish marina profitability. Since any potential reduction in real estate taxes is uncertain, the current real estate tax liability should be used within the report. If the property is overassessed, the appraiser can recommend a tax appeal.

Riparian Rights and Grants

To determine the annual riparian rights payment and the current status of payment, the appraiser needs to call the appropriate government agency. The annual riparian rights fees are an expense when submerged land is leased from the state. In the case of a riparian grant, the tax assessor will usually apply real estate taxes to the submerged land. This is also an expense.

Licenses

The various governmental bodies may require annual license fees to be paid.

Insurance

In general, insurance is more expensive for marinas than for many types of commercial real estate. The nature of marina activity creates more potential hazards than activities at most commercial properties.

Few marina owners have comprehensive liability maintenance programs. Several types of insurance are commonly chosen by marina owners. General liability covers items such as fires, simple on-site accidents, damage to the marina from storms or theft, and negligence claims. Frequently, marina owners also choose a policy covering damage and loss of boats.

The more products and services that a marina offers, the more insurance the marina needs. Marinas need additional insurance for dry rack storage or the use of equipment such as forklifts. If the marina sells, leases, or rents products such as jet skis, new or used boats, motors, or other boat equipment, it must purchase insurance for product liability. Similarly, a marina that offers repairs must purchase service liability insurance.

Utilities

Utility expenses vary significantly from marina to marina, depending upon the quality of management and the size of the facility. The appraiser should obtain operating and expense information for several years to determine trends and compare them to national averages found in the International Marina Institute's *Financial & Operational Benchmark Study for Marina Operators.*[7] (An example of this analysis is found in the case study.)

Chapter 2 listed the wide range of utilities common to marinas. Electricity, water, gas, and sanitary sewer are usually provided to the slips and to the buildings. Buildings receive telephone services, and cable television may be offered to slips

---

7. *Financial & Operational Benchmark Study for Marina Operators* (Wickford, Rhode Island: International Marina Institute, 1996). Equivalent information for the top 25% of marinas participating in the study are also available from this publication.

with houseboat rentals. Some marinas provide gasoline docks for refueling. Fuel, oil, and utilities are some of the largest expenses incurred at a marina.

Advertising

In most cases, advertising is a relatively small operational cost for a marina because most marinas are located in good locations. Billboards and street signs provide adequate advertising for marinas near major roadways. Marinas near ocean access points need little advertising because they draw patrons due to their superior locations. Yacht clubs and other prestigious establishments often have waiting lists because of their popularity.

On the other hand, advertising costs can be high for marinas in poor physical condition or those located in poor locations. Poor locations include areas far from ocean access points or areas that are time-consuming to reach. By determining the market position of the marina, the appraiser can decide the amount that should be allotted for advertising.

Management

Throughout this book, the importance of knowledgeable management has been emphasized. Managing a marina requires more than general real estate knowledge. The manager needs specialized insight about the business and its financial flows.

The appraiser must be aware that applying the same management ratios used for most commercial properties will understate the true costs of management. Management expenses can be difficult to gauge because many marina businesses are mom-and-pop family businesses where the owners do most or all of the work. For this reason, the *Financial & Operational Benchmark Study for Marina Operators* does not list management as a single line item. However, there is a line item for owner's profit. The best way to determine management expenses is to survey local marina management companies. If marina management companies are not covering the labor requirements for all operations and businesses at the marina, the appraiser may also have to calculate the number and type of marina personnel necessary to run the facility

Legal and Professional Fees

Under normal conditions, legal and professional fees are minor expenses. These fees vary with the amount of activity present at the marina and the types of business transactions. If there are environmental problems or stigma associated with the facility, these problems will result in higher than normal legal and professional fees.

Labor

Labor is one of the largest expenses of a marina. The amounts and types of labor depend upon the types of businesses present and the size of the facility. Labor fees usually include the salary of a dockmaster, who oversees operations at the marina, and dock assistants, who provide ancillary services such as boat washing, weatherproofing, and other basic functions. Other employees are foremen and general laborers who supply ancillary services such as bottom painting and winterizing.

Repairs and Maintenance

Repairs and maintenance may be needed on the buildings at the marina, the travel lifts, and the docks. Most marina operators do not allow annual reserves for replacement. For this reason, repair and maintenance figures may be distorted for several years as major repairs are undertaken.

Reserves for Replacement

As mentioned, most marina owners do not add a line item in their cash flow statements for reserves. Reserves are appropriate because the marina buildings have a variety of short-lived components which depreciate over time. Cranes and other personal property used in the daily operation of the business wear out over time. The piers, anchor systems, bulkheads, and other marina specific items have a fairly long life, but they need repair if they are rotting or damaged. The appraiser should make an allocation for reserves to account for these depreciable items.

Other Expenses

A variety of incurred expenses, such as office supplies, telephone costs, postage, etc., may be classified as other expenses.

Operating Expense Ratio Analysis

As part of the income capitalization approach, the appraiser can also examine the total operating expense ratio of the subject and compare it to the "Common-Sized Income Statement" featured in the *Financial & Operational Benchmark Study for Marina Operators*. Operating expense ratios derived from comparable sales provide an additional indicator. When the operating expense ratio is unusually high or low, this may be an indication that extenuating circumstances created the difference. The appraiser should inquire with management about this difference, especially when one or two income and expense items vary significantly from the market or the survey. Examples of national ratios, income statements, and balance sheets are presented at the end of Chapter 4.

Other Ratios and Indicators

In conjunction with the operating expense ratio, the appraiser may also want to examine a number of other ratios and performance indicators. Analyzing changes in the balance sheet can provide insight into financing practices, capital expenditure patterns, and liquidity. Financial ratios such as the current ratio, quick ratio, debt to equity ratio, and other indicators can provide useful information to help gauge changes in operations and management performance. The *Financial & Operational Benchmark Study for Marina Operators* has in-depth analyses of these ratios that can be used for comparison.

Income Capitalization Approach Techniques

The choice of the technique to be used within the income capitalization approach depends upon the preferences of the appraiser, the nature of the assignment, and the type of information that is available from the market. The appraiser should use a technique which accurately models the thinking of a typical marina purchaser in the local market.

Direct capitalization, discounted cash flow analysis, mortgage equity, and band of investment are possible choices. If the appraiser can derive overall capitalization rates, direct capitalization should be given preference over the other techniques (although the other techniques can be used as additional support). Direct capitalization is the most market-oriented technique.

Selection of Overall Capitalization Rate or Discount Rate

Marinas are highly specialized and require more intense management than many other types of commercial real estate. These factors should be reflected in the selection of the capitalization rate or discount rate. Other factors to consider include location, supply and demand, the condition of the improvements and the docks, and other pertinent issues. The appropriate rate will have a risk and management premium higher than the premium for riskless, easily managed forms of real estate such as the investment-grade real estate quoted in real estate investor surveys.

Tables 4.1 through 4.4 show a variety of statistics compiled by the International Marina Institute, Appraisers may find these statistics useful for comparison, as they select rates and techniques to use for the income capitalization approach.

TABLE 4.1 FINANCIAL RATIOS—ALL MARINAS BY SALES			
Based on 1995 Financial Data			
Median Financial Ratios	**$0 to $750,000**	**$750,001 to $1.5 Million**	**$1.5 Million to $15.5 Million**
Liquidity			
Current ratio	1.77	1.37	1.41
Quick ratio	1.11	0.39	0.46
Safety			
Debt to equity	0.98	0.94	1.40
Net sales to equity	1.18	1.30	0.76
Net profit to equity	0.03	0.06	0.04
Net fixed assets to equity	120.7%	100.7%	103.7%
Profitability			
Gross profit margin	73.6%	59.8%	71.9%
Operating profit margin	10.6%	8.8%	15.5%
Pretax profit margin	0.9%	5.1%	2.0%
Net profit margin	0.9%	5.3%	2.0%
Asset Utilization			
Sales to assets	0.71	0.67	0.65
Sales to net fixed assets	0.99	1.32	1.09
Return on Equity & Assets			
Operating return on equity	6.0%	8.1%	10.8%
Operating return on assets	4.7%	4.8%	7.7%
Pretax profit return on equity	1.6%	5.7%	6.0%
			(continued)

TABLE 4.1 FINANCIAL RATIOS—ALL MARINAS BY SALES (CONTINUED)

Based on 1995 Financial Data

Median Financial Ratios	$0 to $750,000	$750,001 to $1.5 Million	$1.5 Million to $15.5 Million
Return on Equity & Assets (continued)			
Pretax profit return on assets	1.9%	2.7%	1.3%
Working Capital Management			
Sales to working capital	1.99	2.57	0.68
Accounts receivable days	18	18	19
A/P turnover days	19	22	36
Inventory turnover days	29	38	40
Operations Management			
Avg. sales per occ. slip*	$1,632	$1,965	$5,761
Avg. sales per dry storage unit**	$729	$978	$1,498
Sales per employee high season	$55,415	$65,803	$78,893
Sales per employee low season	$108,868	$128,478	$137,633
Other			
Avg. total sales	$528,218	$1,011,966	$2,711,891
Operating expense percentage	48.0%	56.7%	56.6%
Avg. interest expense (% of sales)	5.6%	3.7%	5.8%
Number of respondents	19	23	23

* Dockage sales/(total # slips × occ. rate)
** Assumes 100% occupancy of dry storage units

Courtesy of the International Marina Institute, *Financial & Operational Benchmark Study for Marina Operators* (Wickford, Rhode Island: International Marina Institute, 1996). Equivalent information for the top 25% of marinas participating in the study are also available from this publication.

TABLE 4.2 FINANCIAL RATIOS—ALL MARINAS BY REGION

Based on 1995 Financial Data

Median Financial Ratios	Region I	Region II	Region III	Region IV	Region V
Solvency and Liquidity					
Current ratio	1.41	1.28	0.63	1.02	1.75
Quick ratio	0.43	0.55	0.34	0.73	0.57
Safety					
Debt to equity	0.87	1.30	0.11	0.06	0.51
Net sales to equity	0.86	1.08	0.52	0.36	0.66
Net profit to equity	0.03	0.06	0.01	0.06	0.04
Net fixed assets to equity	88.4%	164.7%	75.5%	72.6%	110.8%
Profitability					
Gross profit margin	59.8%	59.9%	60.3%	55.0%	43.1%
Operating profit margin	7.4%	14.8%	4.4%	30.8%	16.8%
Pretax profit margin	0.9%	2.0%	-0.4%	34.4%	5.5%
Net profit margin	2.8%	2.0%	-0.4%	20.7%	4.8%
Asset Utilization					
Sales to assets	0.72	0.51	0.08	0.36	0.03
Sales to net fixed assets	1.25	0.78	0.08	0.82	0.04
Return on Equity & Assets					
Operating return on equity	4.2%	16.7%	3.3%	3.3%	13.5%
Operating return on assets	4.8%	7.1%	5.2%	17.6%	11.3%
Pretax profit return on equity	3.3%	4.5%	1.0%	5.7%	31.5%
Pretax profit return on assets	1.2%	1.4%	-1.6%	16.3%	5.4%
Working Capital Management					
Sales to working capital	0.54	0.41	0.00	3.14	0.34
Accounts receivable days	21	27	15	25	10
A/P turnover days	16	22	39	15	36
Inventory turnover days	80	35	37	26	31

(continued)

TABLE 4.2 FINANCIAL RATIOS—ALL MARINAS BY REGION (CONTINUED)

Based on 1995 Financial Data

Median Financial Ratios	Region I	Region II	Region III	Region IV	Region V
Operations Management					
Avg. sales per slip*	$2,223	$1,131	$2,228	$921	$1,866
Avg. sales per dry storage unit**	$758	$425	$1,245	$480	$1,574
Sales per employee high season	$72,402	$63,556	$71,805	$33,280	$80,620
Sales per employee low season	$146,057	$132,825	$108,985	$87,712	$137,084
Other					
Avg. total sales	$894,776	$780,735	$533,805	$961,108	$779,119
Operating expense percentage	47.5%	44.9%	51.3%	47.5%	63.9%
Avg. interest expense (% of sales)	2.6%	4.9%	6.9%	4.7%	2.9%
Number of respondents	21	14	22	5	17

* Dockage sales/(total # slips × occ. rate)
** Assumes 100% occupancy of dry storage units

Courtesy of the International Marina Institute, *Financial & Operational Benchmark Study for Marina Operators* (Wickford, Rhode Island: International Marina Institute, 1996). Equivalent information for the top 25% of marinas participating in the study are also available from this publication.

TABLE 4.3 COMMON-SIZED INCOME STATEMENT

All Marinas—By Sales

Averages	$0 to $750,000	$750,001 to $1.5 Million	$1.5 Million to $15.5 Million
Revenues			
Dockage	38.0%	31.2%	41.0%
Dry storage/launch	9.0%	6.9%	9.1%
Restaurant/concessions	0.4%	4.6%	1.8%
Fuel/oil	12.2%	16.9%	11.9%
Bait/tackle/sundries	7.8%	3.7%	4.1%
Parking/storage shed	0.4%	0.2%	0.2%
Haul out/repairs	17.6%	18.5%	14.6%
Boat launch revenue	1.2%	0.4%	0.4%
All other revenue	13.3%	12.4%	16.9%
Total revenue	100.0%	100.0%	100.0%
Cost of Goods Sold			
Direct dockage costs	3.8%	1.1%	2.3%
Direct labor costs	11.1%	9.8%	5.6%
Direct costs of merchandise sold	16.7%	26.2%	16.8%
All other direct costs	2.3%	1.6%	4.4%
Total cost of revenue	33.9%	40.5%	29.2%
Gross profit	66.1%	59.5%	70.8%
Operating Expenses			
Advertising	0.9%	1.4%	1.1%
Automobile	0.3%	0.4%	0.3%
Bad debts	0.2%	0.2%	0.3%
Depreciation/amortization	8.9%	5.8%	7.2%
Dues & subscriptions	0.2%	0.2%	0.1%
Employee benefits/taxes	3.2%	2.7%	2.7%
Equipment rent/lease	0.5%	0.3%	0.2%
Insurance—employees	1.0%	0.9%	0.9%
Insurance—liability	3.3%	3.1%	3.0%
Legal & accounting	0.8%	0.9%	1.1%
Office supplies	0.7%	0.8%	0.7%
Owner's compensation	5.3%	2.5%	1.4%
Rent & lease	5.2%	3.3%	5.2%

(continued)

TABLE 4.3 COMMON-SIZED INCOME STATEMENT (CONTINUED)

All Marinas—By Sales

Averages	$0 to $750,000	$750,001 to $1.5 Million	$1.5 Million to $15.5 Million
Operating Expenses (continued)			
Repairs & maintenance	3.3%	3.2%	3.5%
Salaries & wages	10.8%	11.7%	12.3%
Telephone/communication	0.7%	0.5%	0.4%
Travel & entertainment	0.4%	0.5%	0.2%
Utilities	3.1%	3.4%	2.9%
Other expenses	6.4%	3.4%	8.0%
Total operating expenses	55.2%	45.2%	51.5%
Operating profit (loss)	10.9%	14.3%	19.3%
Other Income/Expense			
Other income	2.4%	3.3%	6.1%
Other expense (-)	-3.1%	-1.6%	-17.1%
Interest expense (-)	-5.1%	-8.0%	-26.3%
Total other inc./exp.	-5.8%	-2.7%	-37.2%
Profit before tax	5.1%	11.0%	12.1%
Income taxes (-)	0.0%	0.6%	0.8%
Net profit after tax	5.1%	10.3%	11.2%

Note: Due to rounding and inconsistent detail on survey form, not all subtotals sum accordingly.

Courtesy of the International Marina Institute, *Financial & Operational Benchmark Study for Marina Operators* (Wickford, Rhode Island: International Marina Institute, 1996). Equivalent information for the top 25% of marinas participating in the study are also available from this publication.

TABLE 4.4 COMMON-SIZED BALANCE SHEET—ALL MARINAS BY SALES

Expressed as a % of total assets

Averages	$0 to $750,000	$750,001 to $1.5 Million	$1.5 Million to $15.5 Million
Assets			
Cash	5.2%	5.2%	3.1%
Short-term securities	2.8%	0.3%	5.0%
Receivables—net	1.8%	2.2%	3.0%
Inventory	1.8%	8.1%	1.7%
Other current assets	4.3%	1.6%	4.8%
Total current assets	15.9%	17.5%	18.2%
Property	77.3%	76.6%	60.7%
Furniture & fixtures	3.9%	1.3%	0.8%
Vehicles/mach./equip.	11.5%	13.9%	13.9%
Other fixed assets	16.2%	7.7%	7.8%
Total gross fixed assets	112.3%	99.5%	96.3%
Less: accum. depreciation (-)	-30.6%	-31.3%	-26.8%
Net fixed assets	81.8%	68.2%	76.6%
Other non-current assets	2.3%	14.3%	5.2%
Total non-current assets	2.3%	14.3%	5.2%
Total assets	100.0%	100.0%	100.0%
Liabilities			
Notes payable	14.2%	6.0%	5.9%
Current portion of long-term debt	1.0%	1.6%	3.6%
Accounts payable	1.0%	1.1%	1.2%
Accruals	0.8%	0.8%	1.1%
Taxes payable	0.2%	0.2%	0.2%
All other current liabilities	1.1%	5.0%	2.2%
Total current liabilities	20.5%	14.8%	14.5%
Long-term debt	29.5%	38.3%	52.9%
Other non-current liabilities	0.0%	1.9%	1.4%
Total non-current liab.	29.5%	40.2%	54.2%
Total liabilities	50.0%	54.9%	66.1%
Equity (net worth)	50.0%	45.1%	33.9%
Total liabilities & equity	100.0%	100.0%	100.0%

Courtesy of the International Marina Institute, *Financial & Operational Benchmark Study for Marina Operators* (Wickford, Rhode Island: International Marina Institute, 1996). Equivalent information for the top 25% of marinas participating in the study are also available from this publication.

SUMMARY

The three traditional approaches to value—cost approach, sales comparison approach, and income approach—can be used to value marinas. The cost approach is rarely used for existing properties. It can be a reliable valuation method for a proposed marina if it can be completed without compromising essential elements. To successfully use the cost approach, the appraiser must complete two steps. These include making a detailed inventory of all the improvements on the site (including their dimensions) and accurately deriving land value.

To apply the sales comparison approach to the valuation of a marina, the appraiser should follow the same general procedures used for any other property. The appraiser must consider items specific to marinas such as units of measurement, upland and riparian rights, utilities and services, and buildings and conditions.

For an operating marina, the income capitalization approach is the preferred approach to use. To complete this approach, the appraiser must obtain information about the income and expenses for each area of the marina. In addition to examining the individual expenses, the appraiser may also want to look at the operating expense ratio of the subject, financial ratios, and changes in the balance sheet. Since marinas are more specialized than other types of commercial real estate, the overall capitalization rate or discount rate will have a risk and management premium higher than the premium for riskless, easily managed forms of real estate.

CHAPTER FIVE

Special Valuation Issues

Since marinas are specialized properties, unique issues can confront the appraiser during the valuation process. Some of these issues include business elements, marina boat handling equipment, dockominium and other ownership interests, and lake marinas.

Business elements, boat handling equipment, and dockominium interests were introduced in Chapter 2. This chapter expands the explanations given in Chapter 2 and introduces information about these elements relating to valuation.

BUSINESS ELEMENTS

Like many types of commercial property, the income from a marina property is often generated by some type of business activity. Appraising small marinas may involve elements of quasi-business value where goodwill is absent. Appraisals of large marinas may involve elements of complete business value.

Gasoline sales to the slips and repairs to the boats are the most common forms of business activities at large and small marinas. These services make the marina more marketable and usually bring about a higher occupancy rate. Small operations must offer these two services to be competitive in the market. When these services are present at a small operation, the appraiser should state in the report that elements of a quasi-business nature are present in the marina.

If a restaurant, boat sale showroom, or other similar commercial establishment is present, the appraiser may need to include a complete business value in the report. With these types of ventures, the business elements of a marina can far exceed the value as real estate only. Boat dealerships and repair facilities often have gross and net incomes far greater than that of the marina. Of course, the presence of one or more of these operations can cause greater demand for slips, translating into a lower vacancy rate, more slip revenue, and a higher value for the marina component taken alone. The appraiser must be aware of the income generated by the business elements in the marina and value it accordingly.

BOAT HANDLING EQUIPMENT

In order to remain competitive with other marinas, most marinas own or lease some type of boat handling equipment. Some types of small equipment do not contribute much to the overall value of the marina. Larger, more expensive equipment may impact value. Examples of small and large equipment and brief descriptions are provided.

Mobile or travel crane. These large cranes are mounted on trucks and are easily operable where land is at a premium. The cranes have telescoping beams that can pick up medium-sized boats. Marinas often purchase used mobile cranes from the construction industry.

Forklift trucks. These vehicles are a cross between a forklift and a truck, allowing boats to be picked up between their large forks. The trucks are versatile enough to pick up smaller items such as engines and other heavy objects needing repair.

Mobile saddle lift trucks. These trucks have very large wheels and heavy duty canvas slings. The truck lifts the boat in and out of the water using the large slings.

Stationary lifts. Stationary lifts are affixed to the dock area. They are easy to use, require little maintenance, and can lift and lower boats directly into the water.

Stacking cranes. Stacking cranes are large, heavy-duty cranes that usually operate on tracks.

Hydraulic trailers. Recently, hydraulic trailers have been used for boat handling. These are oversized trailers which use hydraulically operated arms to lift the boat hull into the trailer. The trailer and boat can then be towed to any location.

Stationary lifts and forklift trucks do not contribute much to the overall value of the marina. However, large equipment such as stacking cranes do affect value. Obviously, it is more efficient to have multiple types of boat hauling equipment so a large, expensive machine is not used to move a small boat. When preparing the report, the appraiser should include any equipment needed to run the marina and maintain its income. The report should prominently state whether the equipment was included in the appraisal; if it was not included, the report should state the implicit effects of not including the equipment in the final value conclusion.

Along with boat repairs, gasoline sales are the most common business activities at marinas. Every other slip at this floating dock is fitted with a gasoline pump.

Boat handling equipment like this travel crane can either be leased or owned by a marina operator. Marinas often purchase mobile cranes from the construction industry.

A problem can arise with valuation when boat hauling equipment is leased. When leased equipment is included in the appraiser's reconstructed operating statement, it is implicitly considered within the income approach valuation. However, it may not have been considered in the sales comparison approach unless the marinas selected for comparison have similar equipment.

OWNERSHIP INTERESTS

There are four types of ownership interests a boater may acquire: a dockominium, a cooperative, a club membership, and a long-term lease. Sometimes, these four interests are collectively referred to as a *dockominium;* however, each has noticeably different marketability and effects on value.

Dockominiums

Ownership of a dockominium, also referred to as an *aquaminium* or *condo-minium,* is defined as private ownership of an individual boat slip. This type of ownership is similar to ownership of a residential condominium. There is usually a dockominium association, and the purchaser owns the water rights of his or her slip in the fee simple estate. The dockominium owner is responsible for his or her proportionate share of the common elements, any riparian rights payments, and real estate taxes.

Dockominiums were popular in the 1980s with most conversion activity occurring between 1984 and 1986. Factors that influence dockominium success include location (good harbors, easy access to the ocean, and a highly developed area), the reputation of the marina, quality of the dock facilities, excess demand for slips, and the relationship between aquaminiums and condominium/townhouse projects. If most of these factors are lacking, dockominium projects may not fare well in the conversion.

Valuation of Dockominiums

The valuation of dockominiums is similar to the valuation of any condominium. The cost approach is rarely used for the same reasons it is not used in the valuation of the marina. The sales comparison and income capitalization approaches are the preferred approaches for valuing dockominiums. The motivation of the typical purchaser helps the appraiser decide which approach is most appropriate.

When the dockominium market is weak and there are few investors, the primary motivation may be owner occupancy. In this type of market, a lack of information and relevance may preclude using the income capitalization approach.

Dockominiums and the Sales Comparison Approach
The sales comparison approach for a dockominium is performed in the same manner as it is performed for most properties. Because there is little difference between dockominium units of similar size, this form of real estate is very amenable to paired sales analysis. The adjustments will differ. Financing may be difficult in "soft markets," and marketing time may be extended. When the appraiser compares the dockominium to other dockominium projects, location, real estate taxes, and operating expenses will need to be adjusted for the different dockominiums. Within the project, the primary differences are slip size, water depth, condition of the docks, and ease of access to the harbor.

Dockominiums and the Income Capitalization Approach
To perform the income capitalization approach for a dockominium, the appraiser follows the same procedures used for valuation of other types of real estate. In most cases, a discounted cash flow (DCF) analysis will not be relevant because a mortgage equity or band of investment approach is a better indication of value and models the thinking of typical investors more accurately than a DCF. To determine the economic rental, the appraiser should consider the same factors used for this step in the sales comparison approach. To select an appropriate discount rate, the appraiser should consider the supply and demand for dockominiums and, if applicable, the higher interest rates charged by financial institutions to fund this type of real estate. Usually, a higher discount rate is evident when dockominiums are compared to more typical real estate investments.

Cooperative
Ownership of a cooperative is the ownership of shares of stock in a corporation. The corporation owns the marina, manages it, and handles daily business decisions. The maintenance fees charged to shareholders include real estate taxes, operating expenses, and a proportionate share of the underlying mortgage of the corporation. The shareholder receives a proprietary lease that has most of the bundle of rights of ownership; however, there are some restrictions and risks:

- In many cooperatives, the board of governors has the right to approve or disapprove potential stock purchasers. This may make it difficult to resell the proportionate shares of a slip.
- If a proprietary lease owner does not pay his or her share of expenses and taxes, the remaining leaseholders are responsible for unpaid expenses. If many leaseholders do not pay, the cooperative corporation may become insolvent and leasehold ownership interests may be jeopardized.
- Potential purchasers may find this type of ownership difficult to understand and trust.

Club Membership

Club membership entitles the boater to an ownership in a club. Facilities include more than just a slip: The use of a clubhouse and other recreational amenities such as tennis courts may be offered. The boater pays an initial up-front membership fee and annual dues thereafter. Although there may be prestige associated with the club and the amenities may be among the best in the area, the boater may be unable to sell or transfer the membership. If it is transferable, there may be restrictions concerning the type of member who is allowed to join the club.

Long-Term Lease

Long-term leases are common arrangements when the marina owner owns the riparian water rights but does not own the submerged land. Since the marina owner does not have title to the land, it cannot be sold and a lease becomes the primary vehicle for conveying some of the bundle of rights. This type of lease is usually created for a 49- to 99-year period and gives the boater the right to use the slip. Upon expiration of the lease, the slip reverts to the seller.

LAKE MARINAS

States with little or no oceanfront or bayfront land may have lake marinas. The Great Lakes region has over 300 of these facilities. During an appraisal, lake marinas are treated much the same as any bay, river, or oceanfront marina. Some physical differences between lake marinas and other types of marinas exist. These differences may require different types of designs, and the appraiser should be familiar with these designs and the reasons for them.

One of the most prominent differences between lake marinas and saltwater marinas is the susceptibility of lake marinas to severe ice damage. In the winter months, freshwater lakes in cold climates freeze much faster than saltwater bodies of water. As the water freezes, it expands, and the marina must be able to withstand the prolonged high pressure caused by the ice. In contrast, ice melts quickly on saltwater bays or oceanfront marinas, exposing them to less pressure. In addition to creating high pressure, the ice on lake marinas gradually shifts laterally causing more damage than it would at a saltwater marina where it melts quickly. The vertical movement of the water beneath the ice can pull or push wooden areas of the marina. Docks may capsize or become twisted because of this movement.

Saltwater marinas are susceptible to different types of weather damage that are as problematic as those caused by ice. Saltwater marinas experience stronger winds, more hurricanes, and higher waves than lake marinas experience. (Exception: The Great Lakes states have greater average wind speeds than many coastal areas.) In coastal areas, the tidal problems accompanying hurricanes can create significant damage to property. As a result of these weather factors, the construction components and physical design of lake marinas and saltwater marinas differ from each other in the following ways:

Type of dock. Lake marinas often have floating docks instead of the fixed docks found at oceanfront marinas. In general, fixed docks are more expensive than floating docks. There is a difference in maintenance expense and longevity associated with the more depreciable floating dock system.

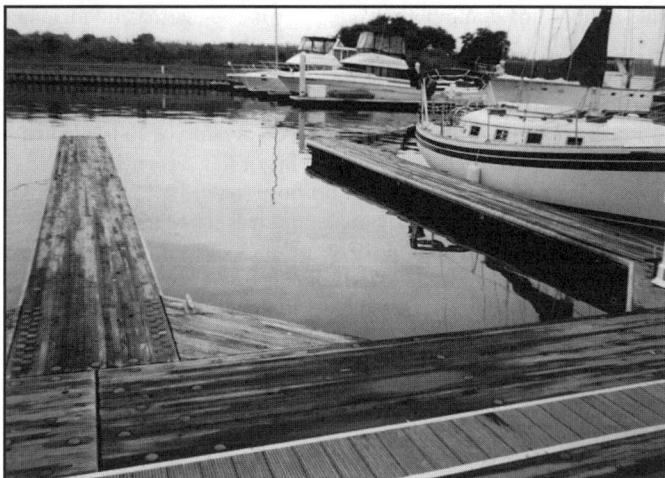

More often used at lake marinas, floating docks like this one are less expensive and more depreciable than the fixed docks found at oceanfront marinas.

Removal of docks and use of deicing equipment. As indicated previously, ice is a problem at many lake marinas during the winter season. Some marinas use equipment to deice the marina, resulting in a higher winter maintenance expense. The appraiser should be aware of this line item when projecting expenses. In areas that suffer from extreme weather conditions, the floating docks of a lake marina may be removed to prevent problems associated with ice. To avoid damage associated with handling and transportation, the floating docks at these marinas must be sturdier and more expensive than average.

Dock anchorage system. The anchorage system for a floating dock is much different than the anchorage system present in a fixed dock system. The degree to which ice typically shifts on a lake, harbor geology, water depths, and wind factors determine the type of system needed for anchoring the floating docks found at lake marinas.

Perforated breakwaters. Breakwaters are prevalent on oceanfront marinas, but it is rarely necessary to have a breakwater on a lake marina. Lake marinas that experience severe weather may have floating breakwaters.

SUMMARY

During the valuation process, the appraiser needs to consider special valuation issues. These include business elements, marina boat handling equipment, dockominium and other ownership interests, and lake marinas.

One of the most important valuation issues is the marina income generated by some type of business activity. Appraising small marinas may involve elements of quasi-business value where goodwill is absent. Appraising large marinas may involve elements of complete business value. The appraiser must be aware of the income generated by the business elements in the marina and value it accordingly.

In order to remain competitive with other marinas, most marinas own or lease some type of boat handling equipment. Small equipment may not contribute much to the overall value of the marina, but larger, more expensive equipment may impact value. When preparing the appraisal report, the appraiser should include any equipment needed to run the marina and maintain its income.

There are four types of ownership interests that affect value: a dockominium, a cooperative, a club membership, and a long-term lease. Ownership of a dockominium is defined as private ownership of an individual boat slip. The purchaser owns the water rights of his or her slip in the fee simple estate and is responsible for his or her proportionate share of the common elements, any riparian rights payments, and real estate taxes. Ownership of a cooperative is the ownership of shares of stock in a corporation. The corporation owns the marina, manages it, and handles daily business decisions. The shareholders pay maintenance fees and receive a proprietary lease that has most of the bundle of rights of ownership, with some restrictions and risks. Club membership entitles the boater to an ownership in a club, which includes not only a slip but also the use of a clubhouse and other recreational amenities. Long-term leases give the boater the right to use the slip. They are usually created for a 49- to 99-year period.

During an appraisal, lake marinas are treated much the same as any bay, river, or oceanfront marina; however, lake marinas may require unique designs because of weather conditions. The appraiser needs to be aware of these designs since they can impact value. Lake marinas often have floating docks instead of the fixed docks found at oceanfront marinas. Floating docks are less expensive than fixed docks and have different types of dock anchorage systems, maintenance expenses, and longevity. Since lake marinas in northern states are more susceptible to ice damage than other marinas, these marinas may have removal docks that are taken down during the winter season. Removable docks are sturdier and more expensive than average. Lake marinas that experience severe weather may have floating breakwaters, but most lake marinas do not have breakwaters.

The consideration of these valuation issues are an important part of the appraisal process. Business elements, boat handling equipment, dockominium and other ownership interests, and lake marinas have a significant impact on the marketability and value of the property.

CHAPTER SIX

Case Study

The following case study is based on an actual appraisal assignment. It uses the concepts presented in the previous chapters, defining the factors and characteristics that make marinas unique from other property types. The report shows all three approaches to value. The reconciliation section correlates the values from each of the three applicable approaches to derive the final indication of value.

Although more complex and involved appraisal assignments sometimes occur, the author selected this case study because it is representative of the typical requirements of a marina appraisal.

SAMPLE APPRAISAL

Appraisal of
A Sample Marina
123 Main Street
Anytown, Maryland

A report prepared for

(Client's first and last name)
(Client's title and organization)
(Client's street)
(Client's city, state, and zip code)

Prepared as of (date of value)

Total Real Estate Services, Inc.
P.O. Box 3371
Crofton, MD 21114
(410) 721-9300

Total Real Estate Services, Inc.
P.O. Box 3371
Crofton, MD 21114
Phone: (410) 721-9300; Fax: (410) 721-5474

(Date of report)

(Client's first and last name)
(Client's title and organization)
(Client's street address)
(Client's city, state, and zip code)

Re: Appraisal of a Sample Marina
    123 Main Street
    Anytown, Maryland

Dear (Client's first and last name):

I have appraised the above captioned property for the purpose of estimating its market value as you requested. It consists of a 110-slip pleasure boat marina, several marina-related buildings, and 3.53 acres of upland and riparian rights. The presence of a small repair operation and the potential for gasoline sales are elements of a quasi-business nature which are included in the valuation.

The appraisal assignment is not based on a requested minimum valuation, a specific valuation, or the approval of a loan. This letter of transmittal is not to be construed as the entire appraisal and is invalid if separated from the attached report. The report, including this letter, describes in detail the method of the appraisal and includes all the data gathered in my investigation. This is a complete, self-contained appraisal.

It is my opinion that an exposure time and a marketing time of one year is reasonable for this property. It is my opinion that the market value of the fee simple estate as of (date of value), subject to the Assumptions and Limiting Conditions contained in the addenda, is

Seven Hundred Fifty Thousand Dollars
($750,000)

Sincerely,
John A. Simpson, MAI, CCIM, IFAS
MD, DC, NJ State Certified
General Real Estate Appraiser

SUMMARY OF SALIENT FACTS

Property type	A pleasure boat marina and accessory buildings
Location	A Sample Marina
	123 Main Street
	Anytown, Anne Arundel County, Maryland
	Block XX, Lot XX

Site

Size	1.0000 acre upland
	<u>2.5317</u> acres riparian
	3.5317 acres total
Topography	Level and open
Frontage	The site has 100 feet of frontage along Main Street
Utilities	All municipal utilities installed
Site improvements	Concrete curbing, street lighting, street sewers, fire hydrant nearby (no concrete sidewalks or on-site sewers)
Utility	Good

Improvements

Type	A 110-slip pleasure boat marina with piers and slips
Buildings	A manager's office, a bait and tackle shop, and a small storage shed
Size	Manager's office - 3,013 SF
	Bait and tackle shop - 480 SF
	Storage shed - 300 SF
Constructed	1956
Purpose of appraisal	To estimate the market value of the fee simple interest as of (date of value), the date of inspection
Marketing and exposure times	One year, assuming no environmental contamination from underground fuel tank
Environmental hazards	A fuel tank is located in-ground. Although I did not see any evidence of leakage, any underground tank is an item of concern. Therefore, *I recommend an environmental audit.*
Wetlands	None, although the site is classified as wetlands
Zoning	B-2, marine commercial. The improvements on the site are allowable and conform to the bulk area requirements.
Outstanding R.E. taxes	None
Prior sale	October 1, 1975, for $100,000
Highest and best use	As improved and currently utilized

SUMMARY OF SALIENT FACTS (continued)

Important conclusions

The office building shows evidence of flood damage.

The subject does not offer on-site land storage for boats because there are no heavy duty boat lifts. This results in *much* less income for the marina and limits its marina slip rentals to relatively small pleasure boats; larger boats, in excess of 30 feet, could not effectively gain water ingress and egress, and these boat owners would not be interested in renting at the subject.

In general, the improvements are in average condition.

Special assumptions and limiting conditions

The presence of a small repair operation and the potential for gasoline sales are elements of a quasi-business nature which are included in the valuation.

I assume that there is no environmental contamination from the underground fuel tank. If contamination is present, it is NOT considered within the value conclusion.

Value estimate

Cost approach	$750,000
Sales comparison approach	$740,000
Income capitalization approach	$750,000
Final estimate of market value	**$750,000**

CERTIFICATION

I certify to the best of my knowledge and belief:

- The statements of fact contained in this report are true and correct. The reported analyses, opinions, and conclusions are limited only by the reported assumptions and limiting conditions and are my personal, unbiased professional analyses, opinions, and conclusions. This appraisal assignment is not based on a requested minimum valuation, specific valuation, or the approval of a loan. My compensation is not contingent upon the reporting of a predetermined value or direction in value that favors the cause of the client, the amount of the value estimate, the attainment of a stipulated result, or the occurrence of a subsequent event.

- I assume no responsibility for matters legal in character, nor do I render any opinion as to the title, which is assumed to be good and marketable. All existing liens and encumbrances have been disregarded, and the property is appraised as though free and clear, under responsible ownership and competent management.

- I have no present or prospective interest in the property that is the subject of this report, and I have no personal interest or bias with respect to the parties involved.

My analyses, opinions, and conclusions were developed, and this report has been prepared in conformance to the requirements of the Appraisal Foundation, the Appraisal Institute's and the National Association of Independent Fee Appraiser's Code of Professional Ethics, the Uniform Standards of Professional Appraisal Practice, and (client's organization) written appraisal guidelines. The use of this report is subject to the requirements of the Appraisal Institute and the National Association of Independent Fee Appraisers relating to review by its duly authorized representatives.

I made a personal inspection of the subject on (date of value). I inspected every sale and rental discussed within the report. No one provided significant professional assistance to the undersigned.

As of the date of this report, John Simpson, MAI, CCIM, IFAS, has completed the requirements of the continuing education program of the Appraisal Institute and the National Association of Independent Fee Appraisers.

My estimate of the market value of the property in the fee simple estate, as of (date of value) and subject to the Assumptions and Limiting Conditions in the addenda, is

Seven Hundred Fifty Thousand Dollars
($750,000)

John A. Simpson, MAI, CCIM, IFAS
MD, DC, NJ SCGREA

1. ASSIGNMENT PARAMETERS

IDENTIFICATION OF THE SUBJECT PROPERTY
The subject of this report consists of 3.53 acres of uplands and riparian rights, several marina buildings, and 110 pleasure boat slips situated at 123 Main Street in Anytown, Anne Arundel County, Maryland. It is commonly known as the Sample Marina and also identified as Block XX, Lot XX on the municipal tax assessment records. The legal description, which is part of the last recorded deed, is in the addenda.

SUBJECT OWNERSHIP/PRIOR SALES INVESTIGATION
The owner of the subject property is Jane Smith, 123 Main Street, Anytown, Maryland. The property was last conveyed on October 1, 1975, from John Doe to Jane Smith for a consideration of $100,000 and recorded in Liber XXXX, Folio XXX in the Anne Arundel County Courthouse. The property is not currently listed for sale or under contract, and there are no options to purchase the property.

FUNCTION AND PURPOSE OF THE APPRAISAL
The purpose of this appraisal is to estimate the subject property's market value as defined within the report. The function of the appraisal is to assist the client or assignee in the underwriting of the risk associated with a commercial loan.

APPRAISAL STANDARDS
This report has been prepared in conformance with the Uniform Standards of Professional Appraisal Practice of the Appraisal Foundation and the written appraisal guidelines of the Appraisal Institute, the National Association of Independent Fee Appraisers, and the client. The presence of a small repair operation and the potential for gasoline sales are elements of a quasi-business nature which are included in the valuation.

DATE OF VALUATION
The property is valued as of (date of value), the date of inspection.

EXPOSURE AND MARKETING TIME ESTIMATES
Based on the data and analyses within this report, it is my opinion that an exposure and marketing time of one year is appropriate for the property, assuming a reasonable listing price.

DEFINITION OF MARKET VALUE
Market value is defined as

> The most probable price which a property should bring in a competitive and open market under all conditions requisite to a fair sale, the buyer and seller each acting prudently, knowledgeably, and assuming the price is not affected by undue stimulus. Implicit in this definition is the consummation of a sale as of a specified date and the passing of title from seller to buyer under conditions whereby:
>
> 1. buyer and seller are typically motivated;
> 2. both parties are well informed or well advised, and acting in what they consider their best interests;
> 3. a reasonable time is allowed for exposure in the open market;
> 4. payment is made in terms of cash in United States dollars or in terms of financial arrangements comparable thereto; and
> 5. the price represents the normal consideration for the property sold

unaffected by special or creative financing or sales concessions granted by anyone associated with the sale.[1]

Fee simple is defined as follows:

> Absolute ownership unencumbered by any other interest or estate, subject only to the limitations imposed by the governmental powers of taxation, eminent domain, police power, and escheat.[2]

ENVIRONMENTAL COMMENTS

Unless it is otherwise noted in this report, I did not observe the existence of hazardous materials, which may or may not be present on the property. I have no knowledge of the existence of such materials in or on the subject property, and I am not qualified to detect such substances.

The presence of hazardous substances such as urea-formaldehyde foam insulation, asbestos, radon, or potentially hazardous materials may affect the value of the property. I based the value estimate on the assumption that there are no such materials on, or in, the property that would cause a loss in value. I assume no responsibility for any such conditions or for any expertise or engineering knowledge required to discover them.

It should be noted that the subject has an underground tank. I am not an environmental inspector. I recommend that the client retain an expert in the field of environmental inspection for any concerns.

EXTENT OF THE APPRAISAL PROCESS

I based the appraisal on the physical inspection of the neighborhood and the subject property. Information came from local multiple listing services and company files and exterior inspections of the comparable sales. I verified data through public records, published demographic data, and sources involved in the sale transactions.

I also gave consideration to the sales comparison, cost, and income capitalization approaches to value. All three approaches were applicable.

2. AREA, COMMUNITY, AND NEIGHBORHOOD

COUNTY DESCRIPTION

Anne Arundel County is located along the Chesapeake Bay, in the Washington, D.C., and Baltimore corridors. The county contains 432 miles of shoreline which is populated with waterfront residences and marinas. Annapolis is the county seat and state capital. Anne Arundel County is known for its boating and water activities, such as crabbing, fishing, sailing, and swimming.

MUNICIPAL DESCRIPTION

Anytown is located in the northern portion of Anne Arundel County. Anytown is a small municipality, covering about a dozen square miles. Like many Chesapeake Bay municipalities, Anytown has become a home for boating enthusiasts, and real estate values reflect this demand. Population is moderately seasonal, although most of the residents live in the town year round. Since Anytown has been completely developed for many decades, there is virtually no change in population from year to year. Route 50 is the primary commercial highway through the area, and a wide variety of commercial roadways are present.

---

1. *Uniform Standards of Professional Appraisal Practice* (Washington, D.C.: The Appraisal Foundation, 1998), 163.
2. Appraisal Institute, *The Appraisal of Real Estate,* 11th ed. (Chicago: Appraisal Institute, 1996), 137.

The neighborhood of the subject is almost exclusively residential in nature. Small, older, shore-like homes are located directly east, across Main Street, from the subject. Other older single-family residences adjoin the northern and southern boundary.

Including the subject, there are several marinas in Anytown (discussed in the income capitalization approach as part of the slip rental survey). Traffic volume is extremely light, typical of residential roadways. Properties are well maintained, and no external obsolescence exists.

3. REAL ESTATE ASSESSMENT AND TAXES

Subject Real Estate Tax Assessment			
	Land	Improvement	Total
Block XX, Lot XX	$525,000	$300,000	$825,000

1997 TAX LIABILITY
The 1997 tax rate is $2.59 per $100 of assessed valuation. The tax liability, based on 40% of the assessed market value above, is $8,547.

ASSESSOR'S MARKET VALUE ESTIMATE VERSUS THE APPRAISER'S FINAL CONCLUDED VALUE
The assessor's estimate of market value can be calculated by dividing the total assessed value of the property by the equalization rate of 40%. The answer is $825,000, or $7,500 per slip. Since the appraised value, $750,000, is similar to the above, a tax appeal is not necessary.

OUTSTANDING TAXES
The tax collector indicated that all real estate taxes and water/sewer charges are current.

4. SITE DESCRIPTION

LOCATION
The subject is situated at 123 Main Street in the Borough of Anytown, Anne Arundel County, Maryland. The property is known as the Sample Marina and is listed as Block XX, Lot XX within Anytown. The addenda include a copy of the legal description of the site from the most recent deed transfer. The site is visually depicted on the tax map.

SHAPE AND TOPOGRAPHY
According to the dimensions from the tax map, the site contains 1 acre of upland. It also has 2.49 acres of riparian rights, a deeded right which allows for the existence and exclusive use of the dock and boat slip areas extending out into the bay. There is 100 feet of frontage along Main Street. The site is level and at street grade.

ACCESS AND VISIBILITY
The marina has direct access to Main Street, good visibility from both sides of the marina, and satisfactory ingress and egress. Main Street is a two-lane, non-divided, local residential road with mild traffic flow.

The marina is not visible from a major roadway, but this is not a concern. Marinas do not have the same need for road visibility as other commercial properties. One of the primary considerations for most boat owners when selecting a marina is the ease of access to prime boating and fishing areas. The subject has direct access to the Chesapeake Bay.

Available Utilities and Services

The site has all municipal utilities installed, namely gas, electricity, water, sanitary sewers, and telephone service. Garbage collection is provided locally. Municipal fire and police protection are readily available.

The marina slips are classified as full-service, although gasoline is not being dispensed at the slips. The slips have water which is an acceptable minimum level under code and 30 amp electrical service. Three gasoline tanks, located on two docks, provide fuel for the boats. Sewer is not provided for the boats.

Although there is a cable box on one of the piers, it does not appear that more than one or two slips have this utility, if any at all.

Flood Zone

Anytown's panel number XXXXXX-XXXXX, dated December XX, 19XX, indicates that virtually all of the subject site is in Flood Zone A5, the second most severe flooding classification within Anytown. Only a small amount of land in the northwest corner of the site is not in a flood zone. (Almost all of Anytown is in some type of flood zone, being located along the Chesapeake Bay.)

My inspection indicated previous flood damage to the office. The site is about five feet above sea level, making it susceptible to major storms and northeasters that can raise the water level high enough to cause flooding. There are no storm sewers on the site or in the area, typical of Anytown. The only way excess water can be removed is through evaporation or run-off into the bay.

Wetlands

The wetlands map indicates that most of the site is classified as wetlands.

Excess Land/Additional Land

Excess land is defined as follows:

> *Improved site.* The land not needed to serve or support the existing improvement.
>
> *Vacant site.* The land not needed to accommodate the site's primary highest and best use.[3]

Excess land is large enough to be subdivided according to the minimum lot size allowed by zoning. Also, the building on the lot must be situated such that the remaining excess land is usable and marketable.

When there is more land than is necessary to support an improvement but not enough to be subdividable, I call this additional land. Obviously, excess land is more valuable because it can be subdivided and sold separately, and it has the added utility of being amenable to the construction of another building. Additional land has less value because an additional building cannot be constructed, and its primary uses are to expand parking or the original building. Excess land is marketable to a typical buyer, but additional land is marketable only to the owner of the original building.

Although the subject's upland contains one acre and the office building contains about 3,000 square feet, most of the remaining site is reserved for parking, boat ingress and egress, and the driveway leading to the street. The accommodation of a reasonable car parking space to boat ratio leaves very little land available for any other purpose. There is no excess or additional land present at the subject.

Easements, Nuisances, and Hazards

The tax map shows that there are no easements on the subject site. Two 10,000 gallon underground fuel tanks supply fuel to boaters by way of fuel lines running underneath the

---

3. Appraisal Institute, *The Dictionary of Real Estate Appraisal,* 3d ed. (Chicago: Appraisal Institute, 1993), 124.

docks, above the waterline. Although I did not notice any problem with the tanks from the ground level, the possibility exists that the tanks could be leaking. *An environmental audit is suggested to determine if any problems are present.* A limiting condition of this report is that there is no such problem present, and the value conclusion is predicated on this fact.

Riparian Rights and/or Grants

Riparian rights are defined as

> The incidental right of an owner of land abutting a body of water to use the water area for piers, boat houses, fishing, boating, navigation, and the right of access for such purposes, limited by public need if on a navigable stream.[4]

With a riparian grant, the owner purchases the land underneath the water, with certain restrictions. It is my understanding that the State of Maryland stopped issuing riparian grants some time ago and most of those that exist are many decades old. No payments need to be made on riparian grants, but annual payments must be made on riparian leases.

Most of the water area assigned to the subject is owned by riparian grant. A relatively small section of water near the northern-most portion of the site is under a riparian lease. An official of a state agency confirmed that the owner is responsible for a $390 yearly payment on this riparian lease. The lease is valid through 2002.

Dredging and Deepwater Status

The subject does not have any deepwater slips, wet berths which are large enough to accommodate the larger yachts or tourist boats. All slips are for pleasure boats only, and they all have a fixed dock design (as opposed to floating docks). A recent dredge report prepared by an engineer indicated that there is no siltation which would preclude access to any slips. This is important due to the difficulties and excessive costs involved with obtaining dredging permits. The addenda include a copy of the dredge report.

On-Site Improvements

The main improvements on the subject land are the buildings described on the following pages. There are wooden pilings and decking and wooden bulkheading where the land meets the water. In addition, there is concrete curbing, a fire hydrant on the site, and a very small amount of on-site asphalt paving (most of the driveway consists of crushed stone). The marina benefits from the street lighting along Main Street.

Storm sewers and concrete sidewalks are not present, but this is typical within Anytown. After several inspections at different times and days, I found the subject's car to boat slip parking ratio to be 0.89 cars per two boat slips. This is in line with the recommendation by the International Marina Institute of allowing one car per two boat slips.

Other Items

The following list about the subject contains items of interest to certain readers. Various third-party organizations require that these items be mentioned in the report.

- The site does not have rail access.
- There are no dry rack storage cradles.
- There is no historical value to the improvements.
- No flammable or radioactive materials are on the premises.
- There are no additional approvals to build.

Conclusion

Assuming that no environmental contamination is present, the subject site has few physical detriments and is well-suited to the current improvements. However, there is a

---

4. Ibid., 312.

noticeable potential for flooding since the site has relatively flat topography and is only five feet above sea level. This is typical for marinas in the area.

5. ZONING ANALYSIS

ALLOWABLE USES
The subject is located in the B-2, marine commercial district. Allowable uses in the B-2 zone include single-family detached dwellings; churches; public parks and recreation facilities; municipal buildings; public and parochial schools; marine services such as dockage, boat hauling, boat building and repairs, and boat sales; marine gasoline stations on bulkheads or docks; outside storage of boats; marine engine sales and repairs; marine supplies and equipment sales; yacht clubs; and automobile parking garages. No conditional uses are listed.

B-2 MARINE COMMERCIAL BULK AREA REQUIREMENTS
As indicated in the table which follows, the subject lot meets all the bulk area requirements.

B-2 Marine Commercial Bulk Area Requirements			
Item	Required	Subject Allocation	Conforming
Minimum lot area	5,000 SF	43,569 SF	Yes
Minimum lot width	50 feet	100 feet	N/A
Minimum lot frontage	20 feet	100 feet	Yes
Minimum side yards	6 feet each side	>6 feet	Yes
Minimum rear depth	25 feet	>25 feet	Yes
Maximum building height	2.5 stories/35 feet	2 stories	Yes
Maximum accessory building height	16 feet	Approx. 10 feet	Yes

ZONING CONCLUSION
The subject is a legal conforming use which meets all the bulk area requirements.

6. DESCRIPTION OF THE SUBJECT'S IMPROVEMENTS

MARINA OFFICE BUILDING
The marina office building is the only occupied and used building. It contains the manager's office, a separate men's and women's shower and lavatory section, and a small laundry area. The building is heated by gas through hot water baseboards.

This one-story building was constructed in 1956. It comprises 3,013 square feet, according to measurements. The exterior is composed of a wooden frame, wood siding, and single-pane glass set in aluminum frames.

The public restroom section of the building contains a men's room with two urinals, two toilets, and a shower, and a women's restroom with two showers, two toilets, and a sink. None of the restrooms are handicap-accessible, but this is typical of older facilities built prior to the Americans with Disabilities Act of 1990. This section is in fair condition.

Adjacent to the public restrooms is the small laundry area containing two washers and two dryers. In addition, there is a 300-sq.-ft. wooden storage unit that has neither heat nor electricity and can be used only for storage.

Other Buildings

In addition to the above, a wooden bait and tackle shop building comprising 480 square feet is located by the docks. It is closed during the winter and not accessible.

Dry Rack Storage

Many marinas use rack storage to warehouse boats. The racks are often two or three levels high and require heavy duty mobile boat lifts to move the boats to the higher elevations. The racks provide a significant source of revenue because they allow additional vertical storage.

The subject does not have any rack storage.

Marina Equipment

Marinas commonly use equipment for day-to-day operations. Marina equipment can include heavy boat lifts, large cranes inside boat repair buildings, company-owned vehicles, and a wide variety of smaller items such as steel jacks and wooden racks for on-site boat storage.

The subject does not have any equipment. Since the subject does not have a boat lift, large boats cannot be hauled into and out of the water. This limits the subject's market to boats under 30 feet in length, resulting in much lower income for the marina. The large upland area remains unused year round. Dry storage on the site cannot be effectively accommodated without a boat lift.

Conclusion

The subject's improvements are in average condition, fully functional, and marketable.

7. HIGHEST AND BEST USE ANALYSIS

Definition

Highest and best use is defined as

> The reasonably probable and legal use of vacant land or an improved property, which is physically possible, appropriately supported, financially feasible, and that results in the highest value.[5]

In the case of improved properties, this analysis is performed separately for the land and the building; and for vacant land, it is performed only for the land. To estimate the highest and best use of the site as though vacant or as improved, a use must pass four consecutive tests:

1. Physically possible—those uses which could be physically accommodated at the site.

2. Legally permissible—those uses legally allowed for the property.

3. Financially feasible—those uses that will produce a positive return.

4. Maximally productive—the only use which produces the highest return.

It should be noted that the subject's small repair operation and potential gasoline sales to the slips create a quasi-business element to the subject which is included within the appraisal.

Highest and Best Use as Vacant
As Vacant—Physically Possible

The site description section indicated that the only detriment to developing the subject site is the fact that it is five feet above sea level. This position contributed to the flooding of the office on at least one occasion. To develop the site, it would be preferable to build

5. Ibid., 171.

improvements on low pilings. Otherwise, it would be prudent to provide containing walls and bring in fill to elevate the land underneath any proposed building. Other than its position above sea level, the site is very developable, as evidenced by the advanced age of the current improvement.

As Vacant—Legally Permitted
The zoning section indicated that the current zoning of the subject property is marine commercial. Marine commercial zoning allows both residential housing and a marina on lots.

The slip vacancy rate survey indicated that marina vacancies are low. This suggests that development of the site as vacant with a marina would be financially feasible. Although a marina fits well with the zoning, the changes in environmental laws during the past two decades would most likely prevent the site from being developed with a marina if it were vacant. The necessary dredging of the slips would create tremendous difficulty because the silt could not be removed without time-consuming government paperwork, multiregulatory agency approval, and an extremely high cost. Installing the bulkheading and piling, if none exists, would be extremely expensive and require specialized conformance to new environmental regulations. (This is the reason that no marina land sales have been approved in well over a decade). However, it is assumed that the bulkheading, wet berths, and planking already exist, and this is not a problem.

As Vacant—Financially Feasible
The competitive slip rental survey, included in the income capitalization approach, indicated that slip occupancy rates are very good during the warm weather months. The subject site is large enough to accommodate dry rack storage, on-site ground storage, and the winterizing of boats. Since the slips are in good condition without siltation problems (as evidenced by the engineer's report) or a significant amount of bulkheading deterioration, a marina use is the only financially feasible use for the site, assuming that the slips are already installed. If the site were vacant without the slips or riparian rights, environmental regulations would make it financially infeasible to develop the site.

As Vacant—Maximally Productive
Assuming that the slips are already installed, developing a manager's office and other improvements is financially feasible. This development would be the maximally productive use of the site. Without the slips, this development is not maximally productive.

Highest and Best Use as Improved
As Improved—Physically Possible
Obviously, the subject's current improvements are physically possible.

As Improved—Legally Permissible
As indicated in the conclusion of the zoning section, the subject's current improvements are an allowable use and meet all bulk area requirements. Any other use would require the permission of the municipality, and it is unlikely that permission would be granted.

As Improved—Financially Feasible
It would not be financially feasible to renovate the marina manager's office; however, it is usable in its current state. Any other use would require significant renovation and not warrant the cost considering the relatively small amount of income from this other use.

As Improved—Maximally Productive
The only financially feasible use for the subject property as improved is its current use; it is therefore the maximally productive use.

Conclusion
The highest and best use as improved conclusion is similar to the highest and best use as vacant conclusion.

8. APPRAISAL PROCESS

VALUATION APPROACHES

An appraiser typically performs one or more of the following approaches to value: the cost approach, the sales comparison approach, and the income capitalization approach.

Sales Comparison Approach

The sales comparison approach adjusts the sale prices of comparable properties to the subject to derive an indication of value.

Cost Approach

With this approach "the value of a property is derived by adding the estimated value of the land to the current cost of constructing a reproduction or replacement for the improvements and then subtracting the amount of depreciation (i.e., deterioration and obsolescence) in the structures from all causes."[6]

Income Capitalization Approach

The income capitalization approach estimates gross revenues, less operating expenses, to develop net operating income which is then capitalized into an indication of value.

APPLICABLE APPROACHES TO VALUE

This report includes all three approaches. The reconciliation section correlates the values from each of the three applicable approaches to derive the final indication of value.

9. THE SALES COMPARISON APPROACH

BRIEF DESCRIPTION OF THE APPROACH

The sales comparison approach is the process of estimating the value of a property by directly comparing it to similar properties that are offered for sale or have recently been sold. To find a range of value indications, the appraiser adjusts units of comparison which differ between a comparable sale and the subject. The best indicator for the approach comes from the range of value indications.

STEPS OF THE APPROACH

The following five steps make up the sales comparison approach:

1. Research the market to obtain information on sale transactions, listings, and offerings similar to the subject.

2. Verify the information by confirming that the data obtained are factually accurate and arm's-length.

3. Select a relative unit of comparison based on the thinking of a typical investor in the marketplace; for marinas, the unit of comparison is price per boat slip.

4. Compare the subject and comparable sale properties using the elements of comparison; adjust the sale price of each comparable appropriately or eliminate the property as a comparable.

5. Reconcile the various value indications produced from the analysis of comparable sales into a single value indication.

---

6. Appraisal Institute, *The Appraisal of Real Estate,* 11th ed. (Chicago: Appraisal Institute, 1996), 90.

Marina Sale One	
Location	Anytown Marina 1 Main Street Anytown, Anne Arundel County, MD
Waterway	Chesapeake
Grantor	John Anderson
Grantee	Arthur's Marina
Block/Lot	1719/242, 242.01, 244, 2478, and 2479
Recorded	Liber 4899, Folio 650
Date of sale	March 1, 1997
Sale price	$850,000
Number of slips	87
Price/Slip	$9,770
Land area	0.69 acre uplands, 0.60 acre riparian rights
Financing	Market

Extracted land value

Sale price		$850,000
Building cost new	$371,560	
Less depreciation	$111,468	
Depreciated value		$260,092
Site improvements	$41,520	
Less depreciation	$8,304	
Depreciated value		$33,216
Bulkheading, piers, and decking cost new	$433,350	
Less depreciation	$173,340	
Depreciated value		$260,010
Land value		$296,682
Land size in SF		30,056
Land value/SF		$9.87

Comments: This is the sale of a marina located on the same bay as the subject. Improvements included two single-family residences containing 2,300 and 1,182 square feet, a bait and tackle shop containing 360 square feet, and a two-story, mixed-use, 1,200-sq.-ft. structure with a luncheonette on the first floor and an apartment on the second. The buildings are in good to very good condition. The marina is a full-service facility. All slips were rented for the upcoming season.

Marina Sale Two	
Location	Anytown Marina 2
	Main Street
	Anytown, Anne Arundel County, MD
Waterway	Chesapeake
Grantor	Arthur Jones
Grantee	Ronald Smith
Block/Lot	1/43 and 1321
Recorded	Liber 5017, Folio 213
Date of sale	October 28, 1996
Sale price	$1,700,000
Number of slips	200
Price/Slip	$8,500
Land area	1.58 acres uplands, 5.3 acres riparian rights
Financing	The buyer assumed a first mortgage for $1,000,000 and gave $400,000 cash at closing; the seller provided a five-year second mortgage for $300,000 on market terms. The parties negotiated the financing after contract, and it did not affect the sale price.

Extracted land value

Sale price		$1,700,000
Building cost new	$0	
Less depreciation	$0	
Depreciated value		$ 0
Site improvements	$49,523	
Less depreciation	$14,857	
Depreciated value		$34,666
Bulkheading, piers, and decking cost new	$1,181,375	
Less depreciation	$236,275	
Depreciated value		$945,100
Land value		$720,234
Land size in SF		68,825
Land value/SF		$10.46

Comments: This is the sale of a marina located in Anytown. The improvements used with this property included a restaurant/banquet building and a 50-unit motel, but they were on a separate lot and were not included in the sale. Essentially, no buildings were included with the transaction. The purchaser renamed the marina and subsequently installed a wooden trailer on the site to handle operations. This is a full-service facility. Even though most of the slips were rented for the upcoming season, the purchaser renovated the slips and landscaped the grounds.

Location	Anytown Marina 3
	Main Street
	Anytown, Anne Arundel County, MD
Waterway	Chesapeake
Grantor	Lee Jackson
Grantee	Honorable Sailing
Block/Lot	1462.09/49 and 49.01
Recorded	Liber 4998, Folio 17
Date of sale	July 30, 1996
Sale price	$1,450,000 total, $1,040,000 for the real estate only
Number of slips	100
Price/Slip	$10,400
Land area	1.274 acres total
Financing	A first mortgage for $700,000 was created at 9% for a 15-year term; financing is equivalent to market.

Extracted land value

Sale price		$1,040,000
Building cost new	$124,320	
Less depreciation	$37,296	
Depreciated value		$87,024
Site improvements	$79,653	
Less depreciation	$23,896	
Depreciated value		$55,757
Bulkheading, piers, and decking cost new	$509,250	
Less depreciation	$152,775	
Depreciated value		$356,475
Land value		$540,744
Land size in SF		55,495
Land value/SF		$9.74

Comments: This is the sale of a marina located in Anytown in the central portion of Anne Arundel County. (The subject is located in the northern section of the county.) The sale price included $165,000 for goodwill, which included a boat franchise; $30,000 for a non-competitive agreement; $15,000 for a wide variety of office equipment; and $200,000 for marina equipment and assorted trade fixtures. The result was a real estate value of $1,040,000. The real estate included the 100-slip full-service marina, a store, and a repair shop. The total gross building area of all structures is 4,144 square feet. Winter storage is provided for about 125 boats.

Marina Sale Four	
Location	Anytown Marina 4
	Main Avenue
	Anytown, Anne Arundel County, MD
Waterway	Chesapeake
Grantor	D & P Associates
Grantee	Gary and Cynthia Cohen
Block/Lot	1323B/1-6, 10-13, 1R, 5R, 6R, 7A, 7R, 8A, 9A, and 10R
Recorded	Liber 4861, Folio 466
Date of sale	September 18, 1995
Sale price	$2,160,000
Number of slips	213
Price/Slip	$10,141
Land area	2.39 acres
Financing	The bank provided five-year financing for $3,612,500 at prime plus 1%.

Extracted land value

Sale price		$2,160,000
Building cost new	$357,360	
Less depreciation	$107,208	
Depreciated value		$250,152
Site improvements	$95,428	
Less depreciation	$28,628	
Depreciated value		$66,800
Bulkheading, piers, and decking cost new	$991,125	
Less depreciation	$198,225	
Depreciated value		$792,900
Land value		$1,050,148
Land size in square feet		104,108
Land value/SF		$10.09

Comments: This is the sale of a marina which included an 800-sq.-ft. repair shop, a 4,400-sq.-ft. maintenance shop, a 1,034-sq.-ft. bathhouse, a 6,934-sq.-ft. showroom, a 5,000-sq.-ft. open shed, and an enclosed storage building. There were gas docks, travel lifts, and other high-quality finishes.

To determine value differences between the subject and the comparables, I analyzed the following items. The magnitude of adjustments for each item depended upon the degree of difference between the subject and the sale being compared; larger differences warranted larger adjustments for an item and vice versa.

Property Rights

Since the degree of property right's influence in a leased fee sale is not easily or precisely defined, I did not make any adjustment for property rights to the sales.

Financing

Financing consisted of seller financing or market financing with third-party lenders. The verification indicated that seller financing did not affect any of the negotiated sale prices. No financing adjustments were necessary.

Conditions of Sale

There were no abnormal conditions, motivational premiums, or discounts. No conditions of sale adjustments were necessary.

Market Conditions (Time of Sale)

There were too few sales that were comparable enough to warrant a paired-sales analysis so I relied upon my judgment for the market condition's adjustment. Between the date of the oldest sale and the date of value, there have been no material differences in supply, demand, and value for marinas. No market condition adjustments were necessary.

Location

Location is a critical factor in determining marina desirability and value. I made adjustments for proximity to a major Atlantic Ocean access point, general desirability of the area, proximity to major roadways, and population density in the area.

Sale One is situated on the bay side of Anytown. The only access to the ocean is provided by Anytown Light. Anytown is fairly convenient to the Route 50 bridge, although it is not as conveniently located to major areas of population. Since Sale One is similar in location to the subject, no adjustment was necessary.

Sale Two is situated in Anytown. There is little difference in density between Sale Two and other sections of Anytown. This marina is situated directly on the Severn River, giving it easier, more direct access to the Atlantic Ocean. As indicated in the slip rental study presented in the income capitalization approach, there is a *significant* difference in rental rates and occupancy levels between a location on the Severn River and a location on an interior bay. For this reason, I adjusted this sale downward by 25% for its superior ocean access.

Sale Three is located in the southern section of Anytown, providing access to the Atlantic Ocean which is similar to the access provided by Sale One. Sale Three has a lower population density than Sale One. There is also significant marina competition between Anytown and Anyplace, resulting in lower occupancy rates and slip rental incomes for Sale Three. However, Sale Three is located along a three-lane commercial road with high traffic volume. Considering all these items, I applied two location adjustments: An upward 15% for an inferior area and more competition, and a downward 15% for location on a commercial highway. The net result was a zero net location adjustment.

Sale Four is situated on the opposite side, as compared to the subject, of the Route 50 bridge. I adjusted it downward by 20% for its superior location and access.

Size of Marina

I adjusted Sales Two and Four upward by 5% for size in consideration of the premise that materially larger marinas have greater economies of scale compared to smaller marinas like the subject (especially when it comes to labor costs). Also, larger marinas have lower prices per slip. Sales One and Three are similar in size and did not require adjustment.

Slip Condition

All of the marina slips were fairly old and in good condition, like the subject. It was not necessary to make adjustments for slip conditions.

Utilities and Facilities

All of the sales are full-service marinas offering the same services as the subject. Repairs and slip gasoline sale operations are generally similar in size to the subject. No adjustments were necessary.

Buildings/Condition

As indicated in the description of the improvements section, the subject contains a small office building in average condition. Sale One had a variety of buildings, and they were all in good to very good condition. I adjusted this sale downward by 25% for its greater number of buildings and their superior condition. Sale Two did not have any buildings. I adjusted it upward by 10% for the lack of a building. Sale Three had a small number of buildings which were in very good condition due to recent renovations. I adjusted it downward by 25%. Sale Four had a variety of buildings in good condition, and all of them were usable. I adjusted this sale downward by 15%.

Land/Size

The subject has an acre of upland which is similar in land size to Sales One, Two, and Three. None of these sales were sufficiently different in land size to warrant adjustments. Sale Four had 2.39 acres. I adjusted Sale Four downward by 15% for its greater land size.

Other Adjustments

The subject and the sales were generally similar in all other factors which influence value. Therefore, no further adjustments were required.

Conclusions

After all adjustments, the marina sales ranged from $5,577 to $7,800 per slip. All four sales received equal weight. In my opinion, $6,700 per slip is reflective of the subject's value. Multiplying the subject's 110 slips by $6,700 per slip results in a value of $737,000, rounded to $740,000.

	Subject	Sale One	Sale Two	Sale Three	Sale Four
Marina Sales Comparison Approach					
Sale price	N/A	$850,000	$1,700,000	$1,040,000	$2,160,000
Number of slips	110	87	200	100	213
Price/slip	—	$9,770	$8,500	$10,400	$10,141
Property rights	Fee simple	Leased fee	Leased fee	Leased fee	Leased fee
Adjustment	—	0.00%	0.00%	0.00%	0.00%
Adjusted price/slip	—	$9,770	$8,500	$10,400	$10,141
Financing terms	Market	Typical	Typical	Typical	Typical
Adjustment	—	0.00%	0.00%	0.00%	0.00%
Adjusted price/slip	—	$9,770	$8,500	$10,400	$10,141
Conditions of sale	Typical	Typical	Typical	Typical	Typical
Adjustment	—	0.00%	0.00%	0.00%	0.00%
Adjusted price/slip	—	$9,770	$8,500	$10,400	$10,141
Market conditions (time)	5/1/97	3/1/97	10/28/96	7/30/96	9/18/95
1995 time adjustment	—	0.00%	0.00%	0.00%	0.00%
1996 time adjustment	—	0.00%	0.00%	0.00%	0.00%
1997 time adjustment	—	0.00%	0.00%	0.00%	0.00%
Total time adjustment	—	0.00%	0.00%	0.00%	0.00%
Adjusted price/slip	—	$9,770	$8,500	$10,400	$10,141
Location	—	0.00%	-25.00%	0.00%	0.00%
Size of marina	—	0.00%	5.00%	0.00%	5.00%
Slip condition	—	0.00%	0.00%	0.00%	0.00%
Utilities and facilities	—	0.00%	0.00%	0.00%	0.00%
Buildings/Condition	—	-25.00%	10.00%	-25.00%	-15.00%
Land size	—	0.00%	0.00%	0.00%	-15.00%
Other	—	0.00%	0.00%	0.00%	0.00%
Total adjustment	—	-25.00%	-10.00%	-25.00%	-45.00%
Adjusted price/slip	—	$7,328	$7,650	$7,800	$5,577

Value range	Low	High
	$613,470	$858,000

Subject size	110 slips
Selected market value	$6,700 per slip
Market value	$737,000
	Rounded to
Market value (rounded)	$740,000

10. THE COST APPROACH

There are five steps involved in the cost approach.

1. Estimate the value of the site as though vacant and available to be developed to its highest and best use.

2. Estimate the replacement cost new of the improvements.

3. Derive the amount of accrued depreciation in the structure(s) from three sources: physical deterioration, functional obsolescence, and external obsolescence.

4. Calculate the replacement cost and depreciation for any accessory buildings and/ or site improvements.

5. Determine the developer's profit for coordinating the enterprise.

Step 1—Land Valuation

As indicated, the first step in the cost approach is the determination of land value. There are six procedures which can be used to value land: sales comparison, allocation, extraction, subdivision development, land residual, and ground rent capitalization. Of these six, the most reliable procedure for marina land valuation is sales comparison, followed by extraction.

Due to environmental, legal, and cost constraints, there are very few new marinas being built in Maryland, and even fewer marinas sell as approved marina sites. The sales comparison technique could not be used for this report because there were no approved marina land sales. Instead, I used the extraction method to estimate the subject's land value.

The Extraction Method

Extraction is defined as follows:

> A method in which land value is extracted from the sale price of an improved property by deducting the value contribution of the improvements, estimated as their depreciated costs. The remaining value represents the value of the land.[7]

In general, extraction is a reliable method for marina land valuations because the percentage of value attributable to the land is quite high compared to the cost of the improvements.

The table below shows the results of applying the extraction method to the four sales presented in the sales comparison approach.

Extraction Method Results from the Sales Comparison Approach				
Item	**Sale One**	**Sale Two**	**Sale Three**	**Sale Four**
Sale price	$850,000	$1,700,000	$1,040,000	$2,160,000
Depreciated value of				
Buildings	$260,092	$0	$87,024	$250,152
Site improvements	$33,216	$34,666	$55,757	$66,800
Bulkheading, piers, and decks	$260,010	$945,100	$356,475	$792,900
Total building value	$553,318	$979,766	$499,256	$1,109,852
Land value	$296,682	$720,234	$540,744	$1,050,148
Land size in SF	30,056	68,825	55,495	104,108
Land value/SF	$9.87	$10.46	$9.74	$10.09
Land value/Sale price	34.90%	42.37%	51.99%	48.62%

7. Ibid., 326.

Selection of Land Value

When I extracted the value of the land from the comparable sales, I arrived at a range. My next task was to consider location and other factors which have a material effect on value.[8] This process required selecting the most comparable property or properties from the range and concluding a land value per square foot.

Sales Two and Four had significant differences in location from the subject; they required material location adjustments in the improved sales comparison approach.

The most comparable properties are Sales One and Three. The values per square foot after extraction are $9.87 and $9.74. I weighted each of these sales equally and selected $9.80 per square foot for the subject. When this amount is multiplied by the subject's one acre of upland (43,560 square feet), the land value is $426,888, rounded to $430,000.

Step 2—Estimating Replacement Cost New of the Improvements

The next step in the cost approach is to estimate the replacement cost new of the improvements. The subject has three types of improvements: buildings, site improvements and marina bulkheading, and decking and piles. The Marshall & Swift valuation service provided estimates for the replacement cost of each of these components. The following table outlines the replacement cost new estimates.

Building Replacement Cost New Estimate		
Direct costs		
Main building	3,013 SF @ $30 per SF	= $ 90,390
Storage shed	300 SF @ $12 per SF	= $ 3,600
Bait and tackle shop	480 SF @ $14.50 per SF	= $ 6,960
Total building costs		$100,950

Step 3—Estimating Accrued Depreciation

Depreciation can come from three sources: physical depreciation, functional obsolescence, and external obsolescence. The subject has no functional or external obsolescence. Based on an inspection of the improvements, I estimate physical depreciation to be 50%. Multiplying 50% by the $100,950 total reproduction cost new gives an answer of $50,475.

Step 4—Replacement Cost and Depreciation of Accessory Buildings and/or Site Improvements

The following table shows the detailed estimate of site improvement costs. To estimate bulkheading costs, I conducted a verbal survey of local engineering companies specializing in the installation of bulkheading.

Site Improvement Cost New and Depreciation Estimate		
Site improvement costs		
Paving, grading, drainage, etc.		$48,000
Bulkheading	80 lineal feet @ $225 per lineal foot	= $18,000
Piling, decking, etc.	110 slips @ $3,400 per slip	= $374,000
Total site improvement costs		$440,000
Less: Accrued depreciation		
Physical depreciation	50% of reproduction cost new	$220,000
Functional obsolescence		$0
External obsolescence		$0
Total building accrued depreciation		$220,000
Depreciated value of buildings		$220,000

---

8. Location and other factors were considered with the improved sales, so it follows that they should also be considered in a land value based on those sales.

Step 5—Estimating Developer's Profit

The estimate of developer's profit and indirect costs such as architectural, engineering, legal services, and other items is 10% of the direct building costs ($100,950) and site improvement costs ($440,000), a total of $54,095.

Cost Approach Summary

The following table summarizes the cost approach calculations. The concluded value from this approach, rounded, is $750,000.

Cost Approach Summary			
Direct costs			
Main building	3,013 SF @ $30 per SF	=	$90,390
Storage shed	300 SF @ $12 per SF	=	$3,600
Bait and tackle shop	480 SF @ $14.50 per SF	=	$6,960
Total building costs			$100,950
Less: Accrued depreciation from all sources			
Physical depreciation	50% of reproduction cost new		$50,475
Functional obsolescence			$0
External obsolescence			$0
Total building accrued depreciation			$50,475
Depreciated value of buildings			$50,475
Site improvement costs			
Paving, grading, drainage, etc.			$48,000
Bulkheading	80 lineal feet @ $225 per lineal foot =		$18,000
Piling, decking, etc.	110 slips @ $3,400 per slip	=	$374,000
Total site improvement costs			$440,000
Less: Accrued depreciation			
Physical depreciation	50% of reproduction cost new		$220,000
Functional obsolescence			$0
External obsolescence			$0
Total site improvements accrued depreciation			$220,000
Depreciated value of site improvements			$220,000
Total depreciated value of the improvements			$270,475
Plus: Entrepreneurial profit/indirect costs			$54,095
Plus: Land value			$426,888
Total cost approach value indication			$751,458
Rounded to			$750,000

11. INCOME CAPITALIZATION APPROACH

INCOME CAPITALIZATION COMPONENTS
To derive a value through the income capitalization approach, it is necessary to estimate five components:

1. Gross income

2. Vacancy and collection loss

3. Operating expenses

4. Net operating income

5. Capitalization rate

Income—Economic Rental Analysis
To determine the economic rental and gross income for the subject, I conducted a slip rental survey of other local marinas. The result of this analysis was a final concluded rental rate which formed the basis of the gross income estimate. The following page shows these rentals.

Subject Slip Rate Discussion
In determining an appropriate slip rental rate, access to the Atlantic Ocean is the key factor. Anytown marinas and this portion of the Anytown Bay are located between two marina markets, essentially creating their own submarket. The marina market to the north, along the Severn River, is obviously the most desirable market because it offers direct access to the Chesapeake Bay and the Atlantic Ocean. The other marina markets include those marinas along and near the Anytown River, in the southern Anytown River area.

Boaters at the marinas in Anytown and the Anytown Bay area must travel through the canal to reach the Chesapeake Bay, and from there negotiate to the ocean. Travel distance is greater at these marinas than at the marinas along the Severn River. During peak summer days, the canal can become backlogged if large ships pass underneath Bridge Avenue (which has a large swinging bridge gate) through the Anytown Canal.

The marinas in the southern Anytown River area are closer to the Anytown Light Channel than the Anytown and Anytown Bay area marinas. In spite of this advantage, it is a comparatively long sailing distance to get to this channel, and waiting times at the channel may be as long as 45 minutes.

In general, marinas along the Severn River command a $30 to $50 per lineal foot premium compared to marinas on the canal, due to easy access to the Atlantic. This amounts to an average of $90 to $100 per lineal foot of boat. The marinas in the southern Anytown River area obtain $50 to $60 per lineal foot on average, much less than slips on the Severn River. It follows that slip rates would be higher for Anytown and Anytown Bay marinas than their southern counterparts, but less than the Severn River marinas. This is because the Anytown marinas are located in more densely developed Anytown, which offers a more convenient boat commute to the ocean via the canal than marinas using the Anytown Light Channel.

Slip Rate Conclusion
The subject has two immediate competitors within easy viewing distance (about one mile or less away). These are the Anytown Yacht Club and Anytown Marina 4.

The Anytown Yacht Club is not comparable to the subject. It is an exclusive private club with a three- to five-year waiting list and high annual dues. The clubs offers over 100 slips, 10 tennis courts, a baseball field, a very large operations building with all kinds of services, and extensive on-site parking. The club does not rent slips to the general public, only to members at discounted rates (the high yearly dues more than offset any lower slip rental rate).

Anytown Marina 4 is the primary competitor for the subject. The rental rates for Anytown Marina 4 are $60 to $70 per lineal foot of boat. (Notice how much less this is than

Marina Slip Rental Survey

Marina	Number of Slips	Summer Vacancy	Winter Vacancy	Waterway	Services	Rent/ Slip	Rent/ Lin. Ft.	Slip Sizes	Comments
Subject Marina Anytown	110	25%	90%	Anytown River	Full service	Summer season only	$84 - $94	18 ft. - 30 ft.	Lacking repair and fueling operations
Anytown Marina 1 South Anytown	247	0%	95%	Anytown River	Full service and restaurant	Summer season only; prices variable depending on boat length	$80 - $105	16 ft. - 60 ft.	Upscale
Anytown Marina 2 South Anytown	207	5%	75%	Anytown River	Full service and restaurant	$1,925 - $7,500 in summer; $375 - $1,300 in winter for wet slips	$96 - $134 in summer; $14 - $25 in winter for wet slips	25 ft. - 50 ft.	Very upscale; dry storage available at nearby Jackson's Marina for $20/lineal foot
Anytown Marina 3 South Anytown	241	5%	90%	Anytown River	Full service	$1,750 - $3,400 in summer; $20/lineal foot in winter	$83 - $100 in summer	25 ft. - 50 ft.	Upscale, although restaurant and other ancillary building services are not available; boats stored on land during winter. Rack storage rates are $58/ft. for under 26 ft. long boats, $68 for over 26 ft. but less than 50 ft.
Anytown Marina 4 Anytown	51 available	0%	60 % - 70%	Anytown Bay	Full service	$1,200 - $3,500	$60 - $70	25 ft. - 55 ft.	Commands good rates on the bay due to extensive boat repair operation and on-site restaurant

the Severn River marinas, a reflection of location differences.) Both the subject and Marina 4 offer full utility services, but Marina 4 has a tremendous advantage because it has travel lifts which can haul large boats in and out of the water, the largest boat repair facility in the area, and an on-site restaurant. As a result, Marina 4 has about 20 slips leased on long-term rentals, mainly to houseboaters, and it is fully occupied during the summer, direct reflections of the desirability of the marina. Clearly, to be competitive, the subject *should* charge less per lineal foot than Marina 4 because a discerning boater would prefer to dock at Marina 4.

To obtain data about rental information for the subject marina, I called Andrew Johnson, the attorney for the owner of the marina, and Debra Foye, the marina manager. Neither would answer questions. Later, I made an anonymous phone call to the manager, and she answered questions about rental rates. Ms. Foye indicated that it would cost $1,700 for an 18-foot boat and $2,300 for a 25-foot boat for the summer season. This translates into $94.44 per lineal foot for the 18-foot boat and $84 per lineal foot for the 25-foot boat.

The subject marina is charging much more per lineal foot than Anytown Marina 4. It appears that the only way the subject marina can maintain a significant occupancy is if Anytown Marina 4 becomes full and potential renters migrate to the subject. Although this probably occurs to some degree, there is one other consideration. The subject's slip rental rates are also *at or above* the rates charged by the Anytown marinas. This explains the subject's greater-than-market 25% vacancy rate in the summer, more than 20% greater than the most vacant marina in the area. Unless both the Anytown 4 Marina and *all* South Anytown River marinas are full, the subject marina will probably not obtain a full occupancy rate due to the subject's high prices.

In addition to the lower rate charged by Anytown Marina 4 and the equal or lesser rates charged by the South Anytown River marinas, all of these marinas have superior boat hauling services. Anytown Marina 4 also has extensive repair yards. Clearly, the subject's rental rates are above market level. Management may charge this amount but the occupancy rate will suffer, especially since boat owners cannot store their boats on dry land during the winter due to the lack of boat hauling equipment available for large boats. (Owners must remove boats altogether from the site.) I selected a rental rate lower than that of the Anytown 4 marina, $60 to $70 per lineal foot, and much lower than that of the Anytown Marinas located on the river, $80 to $134 per lineal foot. A reasonable rate for the subject during the summer is $55 per lineal foot. This lower rate should result in a reasonably high occupancy rate, approximately a 95% occupancy rate. This rate is much higher than the current 75% occupancy rate of the subject at its higher rental rate level.

Slip Season and Dock Storage Capacities

After estimating the occupancy and rental rates, the remaining two steps involved in projecting the gross potential income is to calculate the summer and winter season incomes. Calculating the summer income involves multiplying the economic slip rental income on a per lineal foot basis by the average boat size.

Since the subject does not have hauling capacity, boat sizes are most likely limited to those boats under 30 feet in length. I was unable to obtain information about the boat size that could be accommodated at the slips. In my opinion, an average boat size is 23 feet (between 16 feet and 30 feet). (For reference, the summer boating season lasts from April 1 to October 31, a total of 214 days.)

The winter income for the subject marina is also affected by the lack of boat hauling capacity. Possible income is from long-term leases for houseboats and from owners willing to leave their boats at a slip during the winter. As indicated in the rental survey, very few boaters leave their boat in a wet slip during the winter due to the probability of boat damage and vandalism and the desire to store boats at no cost elsewhere. The high vacancy rates during the winter for all the marinas in the survey reflect these considerations. For these reasons, I did not assign any winter income to the subject; this also accurately reflects the marina's "as is" situation, the underlying criterion for this valuation.

The reconstructed operating statement at the end of this section shows the numerical calculation of the subject's projected income given these parameters.

Gasoline Income—A Quasi-Business Element
Although the subject does not provide gasoline to the slips, it is capable of doing so. To estimate the amount of gasoline and utility income that could be generated at the subject, I queried the marinas in the slip rental survey. Anytown Marina 1 had gasoline sales of $155 per slip; Anytown Marina 2 had sales of $152 per slip; Anytown Marina 3 had sales of $162 per slip; and Anytown Marina 4 had sales of $145 per slip. Since Anytown Marina 4 is the most comparable, I assigned $145 per slip to the subject for slip gasoline income.

Utility Income
The subject provides water and electricity utilities to its customers. The following table outlines those incomes and the estimate which is based on the average of the prior three years.

Utility Income					
1996	1995	1994	Average 1994 - 1996	Marina Institute's Estimate	Appraiser's Estimate
$4,105	$4,896	$3,758	$4,253	Aggregated in Other Income	$4,250

Boat Hauling and Repair Income—A Quasi-Business Element
The subject does not have a repair operation. However, some revenue is generated by boat hauling fees. The table which follows presents the boat hauling income and average over the past three years and the estimate of income, based on the average.

Boat Hauling and Repair Income					
1996	1995	1994	Average 1994 - 1996	Marina Institute's Estimate	Appraiser's Estimate
$5,897	$6,215	$5,789	$5,967	$22,880 (mainly repair income)	$6,000

Other Income
In addition to slip rentals, a marina can generate other types of income. These include year-round dry rack storage income, wet winter slip income, on-site winter storage income, and rental income from one or more buildings (usually a restaurant, residential home, or a bait and tackle shop). Other types of income include power washing income (cleaning the hull of the boat for $2 per lineal foot) and income from an on-site boat showroom.

Of the above forms of income, the subject collects slip rent and utility expense reimbursement. Although I saw two houseboats anchored, I was unable to discover the income generated from them. The subject has virtually no wet slips rented and does not offer dry rack or land boat storage. The site is large enough to accommodate on-site boat storage, but management is not offering this service. The bait and tackle shop is used only during the summer, and I do not know whether it is leased. Considering all these items, I will only assign slip rental and utility reimbursement income to the subject.

Expenses
A marina operation incurs many expenses. These can include advertising, insurance, labor and salaries, legal and accounting, management, office supplies, payroll taxes and benefits, real estate taxes, repairs and maintenance, reserves for replacement, riparian lease payments, telephones and communications, travel and entertainment, utilities, and other miscellaneous expenses. An explanation for each of these expenses appears on the following pages.

Calculating expenses for the subject appraisal involved obtaining the income and expense statements for the prior three years and using a variety of expense ratios prepared by the International Marina Institute. These percentage ratios are based on gross income.[9] (The International Marina Institute compiles and reports operating performance for marinas throughout various regions of the country.)

For simplicity, I took the gross income average over the prior three years, $126,000 rounded, and derived dollar expense estimates for each item for Region 1, the Northeast. The results of these calculations, the prior three years' financing statements, and the reconstructed operating statement appear at the end of this section. The remaining expense discussion shows the reasoning and analysis of each financial statement line item.

Advertising
Like most marina expenses, advertising is seasonal. Most of the advertising budget is allocated in early spring. The following table presents the subject's advertising expense for the past three years, the average expense, the International Marina Institute's ratio estimate, and the appraisal estimate. The appraisal estimate is at the higher end of the range because the subject had the best revenue performance during the years the advertising expense was highest.

Advertising Expense					
1996	1995	1994	Average 1994 - 1996	Marina Institute's Estimate	Appraiser's Estimate
$745	$1,212	$877	$945	$1,134	$1,100

Insurance
Insurance expense includes general liability for the marina improvements, accident insurance, and coverage for hazards. The subject pays higher insurance rates than the averages given by the International Marina Institute. Higher rates are typical of heavily developed urban areas like Anytown. The subject's prior insurance expense history is reflective of the insurance rates charged in this area. The chart shows an average expense within the projections.

Insurance Expense					
1996	1995	1994	Average 1994 - 1996	Marina Institute's Estimate	Appraiser's Estimate
$5,471	$5,278	$5,375	$5,375	$4,158	$5,400

Labor and Salaries
Labor and salary expenses are usually the largest operating expenses, and they are specific to each marina operation. The number and type of services offered have a direct bearing on this expense. It is difficult to compare one facility to another or to use industry averages because of these variations. For this reason and the relatively small amount of variability among the three years' expenses, the subject's performance is the best indicator of expenses. Since the subject has very low personnel turnover and gradually escalating labor expenses, the appraisal estimate is slightly higher than last year's labor and salary expenses.

9. The format of the International Marina Institute's "Common-Sized Income Statement" is different from that reported for the subject. The institute's statement is based on much larger marinas, generally three times the size of the subject. Larger marinas generate much more income, thereby reducing the accuracy of the percentage expenses reported when compared to the subject.

Labor and Salary Expenses					
1996	1995	1994	Average 1994 - 1996	Marina Institute's Estimate	Appraiser's Estimate
$17,524	$16,857	$16,325	$16,902	$27,594	**$18,000**

Legal and Accounting
Accounting expenses for the subject are relatively level, but legal expenses have varied due to a legal claim that was incurred and settled in 1995. The International Marina Institute's low estimate for legal expenses could not be used for this expensive urban area. I based the appraisal estimate on the average legal and accounting expense, which is slightly higher than last year's expenditure.

Legal and Accounting Expenses					
1996	1995	1994	Average 1994 - 1996	Marina Institute's Estimate	Appraiser's Estimate
$2,789	$3,954	$2,189	$2,977	$1,008	**$3,000**

Management
A specialized marina management firm manages the subject. It charges 5% of the income generated by the marina, a typical fee in the local market. The International Marina Institute survey did not consider management fees. The appraisal projection is 5%, rounded. This expense is higher than the expense of the prior three years because of the additional slip gasoline and utility income of the projections.

Office Supplies
The subject's office supply expenses are fairly uniform and generally similar to the estimate of the International Marina Institute. The projections show the average office supply expense.

Office Supply Expense					
1996	1995	1994	Average 1994 - 1996	Marina Institute's Estimate	Appraiser's Estimate
$1,542	$1,388	$1,175	$1,368	$882	**$1,350**

Payroll Taxes and Benefits
Payroll taxes and benefits are slightly above 20% of the labor and salary expense which is typical. These estimates are similar to the International Marina Institute's survey. My estimate is similar to the average.

Payroll Taxes and Benefits Expenses					
1996	1995	1994	Average 1994 - 1996	Marina Institute's Estimate	Appraiser's Estimate
$3,741	$4,285	$4,489	$4,172	$4,032	**$4,200**

Real Estate Taxes
The subject's actual real estate tax liability was used as the basis for this expense.

Repairs and Maintenance
Since reserves for replacement allowances are uncommon in the marina industry, repairs and maintenance expenses tend to be variable. This is the case with the subject which had a low 1994 repairs and maintenance expense but higher expenses in the following two years. The projections use the estimate for 1996, which is slightly higher than the three-year average.

Repairs and Maintenance Expenses					
1996	**1995**	**1994**	**Average 1994 - 1996**	**Marina Institute's Estimate**	**Appraiser's Estimate**
$6,050	$6,380	$4,840	$5,757	$4,158	**$6,000**

Reserves for Replacement
Reserves for replacement are appropriate for the short-lived components of the building. The projected reserves are $0.15 per square foot.

Riparian Lease Payments
The subject has a riparian rights lease which requires an annual payment to the state. (The lease differs from a riparian grant which is fee simple ownership and requires no payment.) An official of the Tidelands Commission indicated that the subject pays $390 annually in riparian rent. This figure appears in the operating statement.

Telephone/Communications
Telephone and communication expenses are fairly uniform. The projection is the average of the prior three years.

Telephone/Communications Expense					
1996	**1995**	**1994**	**Average 1994 - 1996**	**Marina Institute's Estimate**	**Appraiser's Estimate**
$1,211	$1,175	$1,421	$1,269	$882	**$1,200**

Travel and Entertainment
The travel and entertainment expense was atypical in 1996 due to payments for employee attendance at an international conference. The expenses for 1995 and 1994 are more reasonable and are only slightly above the estimate provided by the International Marina Institute. My estimate is an average of these two years.

Travel and Entertainment Expense					
1996	**1995**	**1994**	**Average 1994 - 1996**	**Marina Institute's Estimate**	**Appraiser's Estimate**
$1,274	$687	$799	$920	$504	**$745**

Utilities—Dock
The subject's owner does not offer gasoline utilities at the docks because of the necessary paperwork and liability. The subject could easily offer this utility to slip renters and become a more complete full-service facility. Full-service facilities usually have a higher net operating income. Since the fuel expenses are approximately half of the fuel sale price, I used 50% in the estimate.

Utilities—Office
The office utility expense is relatively uniform. The projection is an average of the expenses.

Office Utilities Expense					
1996	**1995**	**1994**	**Average 1994 - 1996**	**Marina Institute's Estimate**	**Appraiser's Estimate**
$2,789	$3,088	$2,845	$2,907	Part of Dock Utilities	**$2,900**

Other Expenses
Other expenses include a wide variety of miscellaneous charges necessary to operate the marina. This catchall category can be highly variable. The other expense line item in the

International Marina Institute's study is much higher than the subject's other expense category. The projection is an average estimate.

Other Expenses					
1996	1995	1994	Average 1994 - 1996	Marina Institute's Estimate	Appraiser's Estimate
$2,597	$4,017	$4,528	$3,714	$8,064	$3,700

Operating Expense Ratio Analysis

To check the reasonableness of the above expense estimates, an appraiser can look at the operating expense ratio of the subject and compare it to the operating expense ratio of the reconstructed operating statement. The following table details these percentages.

Operating Expense Ratios					
1996	1995	1994	Average 1994 - 1996	Marina Institute's Estimate	Appraiser's Estimate
47.49%	44.76%	44.31%	45.52%	44.70%	46.41%

The operating expense ratio of the reconstructed operating statement is very similar to the subject's performance for the prior three years, the average, and the International Marina Institute's operating expense ratio from its survey. The similarity supports the results of the analysis.

CAPITALIZATION

The final step in the income approach is to capitalize the net operating income into a value. Below is an explanation of the use of the band of investment technique.

Band of Investment

The band of investment technique attempts to model market thinking by dividing the capital investment into its debt and equity components. Simply stated, an investment must achieve a competitive interest rate and risk level for the lender to make the funds available, and an equity investor must receive a competitive cash return in order to invest. The following is a description of these two components.

Debt Component

The debt component of the band of investment consists of the mortgage ratio (M) multiplied by the mortgage constant (R_M). The mortgage ratio is the ratio of the mortgage loan to the amount of the security pledged, commonly referred to as the loan-to-value ratio. The mortgage constant is the ratio of the annual debt service to the principal amount of the mortgage loan at a given period in time, sometimes referred to as the capitalization rate for debt.

Deriving the estimate of these two elements requires analyzing typical market terms. The typical mortgage ratio for the subject area and this type of property is 70% of value. First mortgage rates are 8.50% for fixed commercial loans, and the typical loan term is 25 years, callable in 5 years. Using this data, the mortgage constant is 9.66%.

The debt component is computed as follows:

$$M \times R_M = \text{Debt requirement}$$
$$.70 \times .0966 = 6.76\%$$

Equity Component

The equity component of the band of investment consists of the equity ratio ($1 - M$) multiplied by the equity dividend rate (R_E). The equity ratio is the downpayment on the loan. The equity dividend rate is the relationship between a single year's pretax cash flow expectancy or an annual average of several years pretax cash flow expectancies and the equity investment. In most cases, it is the investor's one-year return on his or her cash investment or downpayment *after* paying all expenses and the debt service.

As stated in the discussion about the debt component, the typical mortgage ratio in the subject market is 70%. Therefore, an investor must contribute the remaining 30% of the investment to complete the transaction. This amount represents the equity ratio.

Due to a lack of owner cooperation, I was unable to derive direct equity dividend rates from the marina sales. However, it is obvious that a high equity dividend rate is appropriate. The success of a marina is dependent on management, and management personnel must have specialized knowledge. For these reasons, marinas are considered risky investments. In my opinion, a 15% equity dividend rate would be necessary to attract equity capital to the subject and this type of investment.

The equity component is computed as follows:

$$(1 - M) \times R_E = \text{Equity requirement}$$
$$.30 \text{ x } .15 = 4.50\%$$

Total Band of Investment Indication
As stated earlier, the band of investment is the sum of the debt and equity components. Therefore, the total capitalization rate indicated by this approach is:

$$M \times R_M = \text{Debt requirement}$$
$$(1 - M) \times R_E = \underline{\text{Equity requirement}}$$
$$\text{Total requirement}$$

$$.70 \times .0966 = \quad 6.76\%$$
$$.30 \times .15 = \quad 4.50\%$$
$$\underline{\quad\quad\quad\quad 11.26\%}$$
$$\text{rounded to } \textbf{11.25\%}$$

Conclusion
The value conclusion from the income capitalization approach is $750,000. The financial performance of the subject over the prior three years, the International Marina Institute's survey of incomes and expenses, the reconstructed operating statement, and the results of the capitalization are found on the following page.

	Actual 1996	Actual 1995	Actual 1994	Marina Institute's Operating Ratios	Appraiser's Operating Statement
Revenues					
Slip income	$118,655	$132,885	$126,421	N/A	$139,150
Slip gasoline income	—	—	—	N/A	$16,000
Slip utility income	$4,105	$4,896	$3,758	N/A	$4,250
Boat handling and repair income	$5,897	$6,215	$5,789	$22,880	$6,000
Other income	—	—	—	N/A	—
Potential gross income	N/A	N/A	N/A	N/A	$165,400
Less: Vacancy and collection loss					
Vacancy	N/A	N/A	N/A	N/A	($6,958)
Collection	N/A	N/A	N/A	N/A	($1,392)
Effective gross income	$128,657	$143,996	$135,968	$126,000	$157,051
Expenses					
Advertising	$745	$1,212	$877	$1,134	$1,100
Insurance	$5,471	$5,278	$5,375	$4,158	$5,400
Labor and salaries	$17,524	$16,857	$16,325	$27,594	$18,000
Legal and accounting	$2,789	$3,954	$2,189	$1,008	$3,000
Management	$6,433	$7,200	$6,798	Not in survey	$7,853
Office supplies	$1,542	$1,388	$1,175	$882	$1,350
Payroll taxes and benefits	$3,741	$4,285	$4,489	$4,032	$4,200
Real estate taxes	$8,547	$8,547	$8,198	Not in survey	$8,550
Repairs and maintenance	$6,050	$6,380	$4,840	$4,158	$6,000
Reserve for replacement	—	—	—	—	$453
Riparian lease	$390	$390	$390	Not in survey	$390
Telephone/communication	$1,211	$1,175	$1,421	$882	$1,250
Travel and entertainment	$1,274	$687	$799	$504	$745
Utilities—dock	—	—	—	$3,906	$8,000
Utilities—office	$2,789	$3,088	$2,845	See above	$2,900
Other expenses	$2,597	$4,017	$4,528	$8,064	$3,700
Total expenses	$61,103	$64,458	$60,249	$56,322	$72,890
Net operating income	$67,554	$79,538	$75,719	$69,678	$84,161
			Capitalized at		**11.25%**
			Value indication:		**$748,096**
			Rounded to		**$750,000**
Operating expense ratios	47.49%	44.76%	44.31%	44.70%	46.41%

12. RECONCILIATION

Value Indications	
Cost approach	$750,000
Sales comparison approach	$740,000
Income capitalization approach	$750,000

PURPOSE OF RECONCILIATION

Reconciliation is the analysis of alternative value conclusions to arrive at a final value estimate. In essence, the purpose of this section is to explain the logic behind the final selection of value.

RECONCILIATION CRITERIA

The reconciliation process assigns merit to one or more of the three approaches to value based on the following criteria:

- Reliability and sufficiency of available data
- Applicability and appropriateness of the approach to the type of property appraised

The Cost Approach

The cost approach was a moderately reliable value indicator for the subject. Using the approach required a large number of estimates, and land value had to be extracted due to a lack of land sales. I gave this approach secondary weight; however, it provided support for the value estimates from the other two approaches.

The Sales Comparison Approach

The sales comparison approach was a reliable market indicator for the subject property. The number of adjustments to the sales were few and the magnitude of these adjustments were not excessive, indicating a strong reliability in the results of the analysis. For these reasons, I gave this approach primary weight.

The Income Capitalization Approach

The income capitalization approach was a reliable value indicator for the subject. The actual performance history of the subject and a large amount of competitive information was available, all of which added to the reliability of this approach. I gave this approach primary weight.

FINAL VALUE CONCLUSION

Based on all of the above, **it is my opinion that the market value on (date of value) of the property in the fee simple estate is $750,000.**

ADDENDA

Assumptions and Limiting Conditions

1. As per client request, this is a complete appraisal prepared in a self-contained report format.

2. I inspected the property and sales, and I accept full responsibility for their description. The analysis, conclusions, and values are solely my product.

3. For buildings constructed prior to 1978, the law requires disclosure of any lead paint problems observed during the inspection. Although lead paint is not as common in commercial properties, areas such as stairwells, hallways, and some interior areas may have, at one time, contained lead paint. There do not appear to be any obvious lead paint problems in the subject. However, I am not a qualified lead paint inspector, and I defer to an expert in this area.

4. The photographs used in this appraisal are digital pictures utilizing imaging technology. I personally inspected the comparables utilized in this report. The photographs used are true and correct representations of the subject and the comparables, and I noted and addressed any defects.

5. I assume no responsibility for the legal description or for legal matters. I assume that the title to the property is good and marketable. Unless I stated otherwise, I appraised the property free and clear of any or all liens or encumbrances.

6. I am not required to give testimony or attendance in court unless arrangements have been previously made.

7. I assume responsible ownership of the property and competent property management.

8. I assume that the legal description furnished is correct. The sketch in this report, if any, is included to assist the reader in visualizing the property only. I made no survey of the property and assume no responsibility in connection with such matters.

9. Unless otherwise noted in my report, I assume that the utilization of the land and improvements (if applicable) is within the boundaries and property lines of the subject and that there is no encroachment or trespass.

10. Neither all, nor any part of the content of the report or copy thereof (including conclusions as to the property value, the identity of the appraiser, professional designation, reference to any professional appraisal organization, or the firm with which the appraiser is connected), shall be used for any purpose by anyone but the client specified in the report; the mortgagee or its successors and assigns; mortgage insurers; consultants; professional appraisal organizations; any state or federally approved financial institution; any department, agency, or instrumentality of the United States; or any state or District of Columbia; without my previous written consent. This information shall not be conveyed by anyone to the public through advertising, public relations, news, sales, or other media, without my written consent and approval.

11. I believe the information identified and contained in this report, as furnished to me by others, is reliable, but I assume no responsibility for its accuracy. I assume that there are no hidden conditions of the property, subsoil, or structures, which would render it more or less valuable. I assume no responsibility for such conditions or for engineering studies which might be required to discover such factors.

12. No environmental impact studies were either requested or made in conjunction with this appraisal, and I reserve the right to alter, amend, revise, or rescind any of the value opinions based upon any subsequent environmental impact studies, research, or investigation.

13. I assume that there is full compliance with all applicable federal, state, and local environmental regulations and laws, unless noncompliance is stated, considered, or

defined in the appraisal report. In this appraisal assignment, I did not consider the existence of potentially hazardous material used in the construction or maintenance of the building, such as the presence of urea-formaldehyde foam insulation, and/or the existence of toxic waste or radon, which may or may not be present on the property and which may or may not affect the value of the property. I based the value estimate on the assumption that there is no such material on or in the property that would cause a loss in value. I do not assume responsibility for any such conditions or for any expertise or engineering knowledge required to discover them. I have no knowledge of the existence of such materials in or on the property, and I am not qualified to detect such substances. If desired, the client should retain an expert in this field.

14. Development of land is subject to various environmental regulations, including regulations regarding wetlands, as well as many other regulations. Any references made to soil types, development capabilities, or the location of wetlands are based on county agricultural soils and national inventory wetland maps. Such maps are useful as a guide only, and their accuracy and reliability cannot be guaranteed. I am not qualified to determine the type or quality of soils or wetland boundaries.

15. This appraisal is made with the understanding that the subject can obtain a negative declaration from the Department of Environmental Protection pursuant to the regulations and requirements of the Environmental Cleanup Responsibility Act of 1983 (ECRA), as amended. This act requires as a precondition of any cessation of operation or the transfer of real property which used or stored regulated hazardous substances, the testing, cleanup, and disposal of any such material. I am not qualified to determine the existence of any such hazardous material and therefore have expressed a value of the subject property as if free and clear of any such substances.

16. In conjunction with the preceding paragraph, I have not been apprised of, nor am I qualified to ascertain, the existence of radon, a radioactive gas which occurs naturally in the soil of certain identified areas. This gas, in concentrated form, has been shown to be detrimental, and its existence would create a negative impact on value. As in the above instance, the value estimate assumes the subject is free and clear of radon gas.

17. Unless a nonconformity has been stated, considered, or defined in the appraisal report, I assume that the subject is in compliance with all applicable zoning, use regulations, and restrictions.

18. I assume that all required licenses, certificates of occupancy, consents, or other legislative or administrative authority from any local, state, or national governmental or private entity or organization have been, or can be, obtained or renewed for any use on which the value estimate in this report is predicated.

19. This appraisal was made in accordance with the rules of the professional ethics and practice of the Appraisal Institute.

20. The Americans with Disabilities Act (ADA) became effective January 26, 1992. Although I may have discussed whether the property conforms to the ADA, I have not made a specific compliance survey and analysis of this property to determine the degree of compliance with the ADA. It is possible that a compliance survey of the property, together with a detailed analysis of the requirements of the ADA, might reveal that the property is not in compliance with one or more of the requirements of the act. If so, this fact could have a negative effect upon the value of the property. Since I am not qualified to do so, I did not consider possible noncompliance with the requirements of the ADA in estimating the value of the property.

21. Acceptance of and/or use of this appraisal report constitutes acceptance of the foregoing Assumptions and Limiting Conditions.

22. **I assume that there is no environmental contamination from the underground fuel tank. If contamination is present, it is *NOT* considered within the value conclusion.**

Bibliography

BOOKS AND SEMINARS

Haddad, Nicholas S. *Appraisal of Marinas, Encyclopedia of Real Estate Appraising,* 3d ed. Englewood Cliffs, N.J.: Prentice-Hall, Inc., 1989.

International Marina Institute. *Marina Design & Engineering Conference Technical Papers.* Wickford, R.I.: International Marina Institute, 1987.

International Marina Institute. *Marina Investment & Appraisal.* Wickford, R.I.: International Marina Institute, 1990.

International Marina Institute. *Dry Rack Marina Handbook.* Wickford, R.I.: International Marina Institute, 1992.

Ross, Neil W., and Paul E. Dodson. *Dockominium: Opportunities & Problems, Proceedings of the 1987 National Dockominium Conference.* Wickford, R.I.: International Marina Institute, 1988.

REPORTS AND PAMPHLETS

International Marina Institute and Management Advisory Services. *Financial & Operational Benchmark Study for Marina Operators.* Wickford, R.I.: International Marina Institute, 1996.

Ross, Neil W., *Auto Parking in Marinas.* North Kingstown, R.I.: International Marina Institute, 1988.